The History of Town Planning in a Dependent State

Louis C. Wassenhoven

The History of Town Planning in a Dependent State

Greece in the 19th Century

Springer

Louis C. Wassenhoven
Professor Emeritus, Department of Urban and Regional Planning
National Technical University of Athens
Athens, Greece

ISBN 978-3-031-96957-7 ISBN 978-3-031-96958-4 (eBook)
https://doi.org/10.1007/978-3-031-96958-4

This Springer imprint is published by the registered company Springer Nature Switzerland AG
The registered company address is: Gewerbestrasse 11, 6330 Cham, Switzerland

To Maria-Evdokia and Alexis

Foreword: The History of Spatial Planning—Does it Help us Plan Space and Place Today?

Emeritus Professor Louis Wassenhoven offers a detailed nineteenth century history of spatial planning in Greece, tracing its development from the birth of the modern Greek nation-state. This was itself a challenging spatial process, starting with its independence in 1830 (by the goodwill of the 'great' European powers at the London Congress), the extension of the territory in 1864 (Ionian Islands—a wedding present by Britain to the new Greek King George), Thessaly in 1881, and Crete, Northern Greece, and some Aegean islands in 1913 after the Balkan Wars. Following WWI Greece 'received' Smyrna (Izmir) and surroundings, but had to cede them again in 1922 after the Greek-Turkish war and the Asia Minor Disaster, followed by massive emigration of Greeks from Asia Minor to Greece. These events mutilated the nascent Greek national territory again and again, fragmented it politically and socially, and heavily mortgaged the opportunities to develop an integrative spatial planning system. In the conclusion of his book, Louis Wassenhoven writes that his argument on the development of town planning in nineteenth century Greece 'has been closely related to the effective subjection of the country to the control of its protecting powers and to the constant flow of imported ideas from the developed European North and West' (p. 155).

To arrive to this conclusion, he uses the criteria Dos Santos and other Dependency authors wield to single out when a nation, region, or locality is dependent on the roles imposed on them by the powers dominating them (Dos Santos 1970). Such roles are usually multifunctional to the development of the dominating nation. Dependency relationships keep a territory under the yoke of international capital, foreign economies, or central regions/cities in the peripheral economy where the dominated territory is located. Typical features of such a relationship include: the location of low value-added activities in the dominated territories ('the periphery'); the concentration of high value-added activities in the dominant territories (the 'centre'); skewed governance systems with low levels of autonomous economic and political decision making, which exclude the dominated territory from access to big money and large-scale projects; and the reproduction of its dependency status by obstructing access to power structures, innovation dynamics, and financial resources. All of this is liberally sauced with foreign ideas and interventions that threaten to smother the territory's own development pathways. Dependency often thrives on the fragility of the dominated territories, which as infant nations with fragmented, low value-added economies grounded in agriculture and crafts—inhabited by communities struggling for survival and the maintenance of their local cultural identity—are easy prey for dominant powers.

Most of these features apply to the infant nation of Greece, which came into being with the Revolution of the 1820s, saw its territory expanded and then curtailed, underwent several forms of foreign-controlled administration and only found some governance stability in the 1920s—the period in which the first real spatial planning law was approved by Parliament.

Yet city and town planning initiatives, as piecemeal as they were, are of an older date. After the Revolution, Spatial Planning in Greece started in the period 1827–1833 under the rule of Governor Count Capodistria whose technical advisors worked out elementary town planning tools such as street and lay out schemes, elementary division into housing quarters, etc. These tools incorporated early tentative concerns about the environment, economic development, and citizen participation. But good intentions do not pave the way to a sustainable territory. As Louis Wassenhoven explains, from the outset 'all renovating efforts aimed at the regeneration of towns are subjected to pressures from a broad variety of political forces' (p. 127), both domestic and foreign, not least from the Greeks of the diaspora. The interplay between these forces materialized into economic initiatives, which mostly incorporated private interests. Politicians and ministers often acted as economic players themselves, many being large landowners with access to significant financial resources. But planning was also under pressure from small landowners who sought to have their interests defended by local political leaders:

> The evidence of these clientelist bonds is provided through the 19th century, by the constant stream of amendments and extensions of town plans. Several of these interventions concern the high-quality urban districts and resorts where land belongs to high class owners, but a good many are located in the urban periphery, where the plans are hastily and carelessly extended (p. 127).

Many conflicting pressures explain why the planning apparatus, even today, adopts technically appropriate tools aimed at making space humane, organized, and sustainable, yet in most cases remains locked in filibustering: laws and regulations are established but not applied; and when implemented, they are often manipulated to serve particular interests or challenged in court and boycotted by landowners and tenants with conflicting claims. These challenges are catalysed by clashing interests among foreign protectors—France, the UK, Russia, and Prussia—each attempting to impose planning approaches ill-suited to a fragile territory under construction; a chronic lack of planning expertise that, even if present, struggles to navigate the political and economic maze of competing external and local actors vying for control over land markets; recurrent financial crises, partly fuelled by hesitant foreign investments and weak governance at national and local levels; ongoing conflicts with Turkey; territorial expansions and contractions; a love-hate relationship of the citizenry with the State. All this explains why the planning apparatus (till today) takes on board technically appropriate tools to make space human, organized, sustainable, but in most of the cases have been locked into filibustering: laws and regulations are ‘set’ but not applied; and when proclaimed, manipulated to the benefit of particular interests, or simply boycotted or challenged in court by landowners and tenants with conflictive interests.

The Twentieth Century in a Historical Mirror

We could say that this institutional unrest in the spatial planning field came to an unstable equilibrium in the early twentieth century[1] with the Presidential Decree of 1923 that brought the urban planning act into force. The legislative documents that followed witness of a more integrative ambition, moving beyond just tools and isolated spatial scales by addressing particular and overarching spatial challenges in a more appropriate way. Extra triggers for spatial planning (city plans e.g.) emerged from the 1975 post-dictatorship constitution and had to bridge a big historical gap:

> Legislation produced during that time [in the 1990s] established a holistic and integrated spatial planning system for the country and also contributed to the production of the new generation of Plans in Greece at all levels (local, regional and national). Innovative amendments in the planning legislation during that period included: (a) the extension of the geographical area covered by Local Plans to the Municipal limits instead of the city limits that had been the practice until then (by Law 2508/1997) and (b) the launching of Spatial Plans at the national and regional levels (by Law 2742/1999) (Papageorgiou 2017, p. 1822).

Efforts were made post-2014 to simplify procedures and reduce implementation times. Institutionally and legally speaking, there was relative progress. Between 2014 and 2020, the number of planning levels was turned back from seven to five, better connections between spatial planning levels and development programming were established, the European Landscape Convention was ratified, ... (Gourgiotis and Tsilimigkas 2024). In general, Greece integrated itself into the European spatial planning and development schemes such as the ESDP and Local Agenda 21. It also took on board strategic spatial planning methods, not the least to invoke exceptions on spatial planning regulations to the benefits of large-scale projects such as in the case of the Olympics (Delladetsimas 2003).

Yet these promising steps forward did not pave the way to a Greek planning paradise. The new procedures and the way they were implemented (or escaped implementation) resonate many of the old problems flagged up by Prof. Wassenhoven in his historical analysis of spatial planning in nineteenth century Greece. Compared to the nineteenth century, the State system has become stronger, but it still has to unroll recovery strategies, especially following the debt crisis of 2009 which could only be solved if Greece met the strict requirements imposed by the EC. That this led to a situation of severe poverty for one third of the Greeks was of less importance to the Troika and the EC. This dramatic socio-economic condition certainly moved

[1] In dynamics in mathematics, an unstable equilibrium is an equilibrium point that can easily be left due to an external factor.

political attention away from so-called minor challenges such as spatial planning and educational policy. Despite (or because of?) living up to the European norms, the planning system remained complex, embedded in an avalanche of laws, degrees and regulations corresponding to different spatial levels, and only integrated among them on paper. Persistent issues include delays in officialization and enactment of so-called approved plans and frameworks, which consequently become outdated; a widespread lack of implementation at the local level—understandably a reaction to the centralized planning system; and, not least, lobbying by private interests aimed at bending plans and regulations to serve their own greed. The shift towards more neoliberal planning, especially to limit regulation, certainly has reinforced such lobbying dynamics.

Lessons to be Learned

Comparing the present to the past can be instructive not only for charting a way forward but also for guiding approaches today. And what holds true for Greece also provides valuable insights for improving spatial planning elsewhere. *A lot can be learned from this distinct history of spatial planning in Europe, its nations and Greece itself. After all, Greece has been historically influenced by numerous European planning doctrines that conditioned its making, but in many respects proved to be incompatible with its local socio-economic dynamics.*

First, not so different from the nineteenth century, spatial planning has become too discursive (words, words, and more words). The spatial planning system in most countries has become a tangle of laws and instructions that are hard to manage on the ground—certainly when they are contradictory. A lesson here is to first look at good practice at all spatial levels and build planning regulations from the bottom-up. Indeed, local planning projects involving constituencies drawing on their own experience often prove to be reasonable and satisfactory in practice.

Hence, *secondly*, contrary to contemporary claims by business leaders and top-down politicians, participatory planning should be solidified, not abandoned. Certainly, some criticisms formulated by practitioners—such as lengthy participation and feedback procedures, an excessive number of appeal points—are ad rem. Procedures can be simplified or contained in time, but should be (re)designed to reveal real needs and grounded solutions.

This leads us to express, *thirdly*, the need for a revival of strategic spatial planning in all its dimensions. Albrechts (2004) makes a solid plea for strategic spatial planning:

> Increasingly, it is being assumed that the solutions to complex problems depend on the ability to combine the creation of strategic visions with short-term actions. The creation of strategic visions implies the design of shared futures, and the development and promotion of common assets. Moreover, all of this requires accountability within a time and budgetary framework and the creation of awareness for the systems of power. Delivering on these new demands implies the development of an adapted strategic planning capacity and a shift in planning

> style in which the stakeholders are becoming more actively involved in the planning process on the basis of a joint definition of the action situation and of the sharing of interests, aims, and relevant knowledge (p. 743).

Strategic spatial planning applies best at the local, municipal, or city level. Essential to it are the inclusion of the concerned citizens, accountability, and the pursuit of collective benefits—while respecting private interests of small and medium income earners. It overcomes a well-known problem that spatial planning is too often disconnected from the economic, social, cultural, and environmental development ambitions, as it is too concerned about 'filling space' as a conglomerate of (spatial) functions.

Fourth, the need for continuity in state agency. This can best be guaranteed by entrusting long-term commitments to a specialized, yet democratically controlled spatial planning administration. It cannot be that public administration is stripped from its main long-lasting civic roles to serve the public interest because of an ideological shift in government. Public administration, decently skilled and staffed, should act as the guardian of the lessons from the past, the custodian not only of cultural heritage but also of the track-record of positive and negative lessons learned. To let public administrators take on that role, they should recur to history in a structured way, situating spatial planning (mal)practice within the time-space conjuncture of each emblematic moment in spatial development.

These are the lessons to be learned from Louis Wassenhoven's historical analysis and perceptions on the Hellenic planning experience.

Frank Moulaert
Emeritus Professor of Spatial Planning
KU Leuven, Belgium

References

1. Albrechts L (2004) Strategic (spatial) planning reexamined. Environment and Planning B: Planning and design 31(5):743–758
2. Delladetsima PM (2003) The Olympic village: a redevelopment marathon in Greater Athens. The globalized city: Economic restructuring and social polarization in European cities 65–90
3. Dos Santos T (1970) The structure of dependence. Am Econ Rev 60(2):231–6
4. Gourgiotis A, Tsilimigkas G (2024) Spatial planning in Greece: framework critical review, challenges and perspectives. Plann Pract Res. https://doi.org/10.1080/02697459.2024.2420510
5. Papageorgiou M (2017) Spatial planning in transition in Greece: a critical overview, European Planning Studies 25(10):1818–1833. https://doi.org/10.1080/09654313.2017.1344194
6. Wassenhoven L (1993) Regional and spatial policy: The European strategy of Greece, international seminar: Greece in Europe: Spatial planning and regional policy towards 2000. YPETHO–YPEHODE–TEE, Athens. (In Greek)

Preface

The book is based on research carried out by the author in Greece, addressed initially to Greek readers. However, a great deal of the research has a definite international interest and its results form the backbone of his new undertaking in English. Although some of the observations included in the initial project are reproduced here, the gist of the author's arguments dwell on issues which are not of a narrow national interest. The book will interest students and historians of town planning, whose area of reference is not only Europe, but also the countries which have been through a state of dependence and external domination, and had to import foreign planning ideas and concepts.

The author tries to embrace a broad spectrum of urban planning ideas, developments and influences which dominated the 19th century planning practice of a country which had just achieved political independence, but was still practically a protectorate of the great powers of the time, often referred to as the Protecting Powers (Great Britain, France, and Russia). Every aspect of domestic policies and developments was directly dictated or indirectly influenced by the choices, policies, laws, and actions of these external forces. It was also influenced by the advice and expert opinions of foreigners or even of Greeks, whose mode of conceptualizing the problems of the dependent country had been deeply affected by their life or education in foreign lands. Looking at this state of affairs more than two centuries after the 1821 Greek War of Independence demands full awareness of these conditions.

When turning his glance to the history of his country, the observer and narrator can adopt a variety of viewpoints. He can limit his account to international or civil wars, military conflicts, political leaders, those who sacrificed their lives, and the invisible heroes. He can pay attention to political feuds and antagonisms, ideological declarations, and social unrest. If he wishes to proceed to a loftier analysis, he has the option to focus on ideas, moral choices, lasting values, culture, or the arts. He can naturally examine the structure of civil society, the organization of the state, the institutions, the country's constitution, the laws, and the administrative apparatus. He can narrow down his interests, and look at government policies for education, social and health care, living conditions, urban development, housing, or, why not, planning for cities and settlements. In all these cases, the last one included, he will fail in his research if he limits the object of his observation to a mere technical analysis. There is no doubt that policy implementation at the level of *praxis* has separate scientific components, but its end-result is entangled in, and dependent on, a complex social and historical setting.

Although the author is by training an urban and regional planner, he realized quickly that to deal with planning in 19th century Greece, in fact in a period up to the 1st World War and of its immediate aftermath, required to broaden his perspective. Although the obvious thing to do was to look at city planning legislation, town plans, decision-making procedures, plan implementation on the ground, and social reactions, there was also a need to explore the town planning laws and experience of powerful European countries, which were transplanted on Greek soil. It was often, and still is, difficult to guess how ideas travelled to Greece, who had actually written an Ordinance or Royal Decree, what was the intellectual load of a non-Greek minister or adviser, or how the findings of foreign research had made their way to the Greek capital of Athens. While studying the contents of the first Greek planning laws or town plans, one can discover innovations, novel ideas, and planning instruments, which were to survive for many decades or found their way into planning legal texts compiled several years later.

At every step what was essential was to rely on a comprehensive knowledge of history, in order to explain why this or that choice was made and why a particular clause was incorporated in Greek laws. Surprising initiatives could be credited to individuals with diverse backgrounds in different disciplines, from engineering and architecture to law and medicine. The rulers of the day, kings or governors, had their own experience and world views, depending on their native countries, previous occupation, or political convictions. They were not of course operating in a vacuum. They had to face private entrenched interests, those of local primates with a great deal of power in Greek traditional society or those entrepreneurs who smelled the prospects of doing business in a new-born state. They also had to accommodate social customs and practices and the application of custom-law inherited from distant centuries.

Planning action could be justified through reference to a long-run strategy, practical need, short-sighted priorities, private choices, social practices, or random contingencies. Examples can be found in a variety of conditions, ranging from the need to settle refugees, rehabilitate a sea-port of trade, develop a site of mineral resources, or reconstruct a city destroyed by fire, to demands to create a foreign colony, improve an unhealthy environment, or design a royal capital city. For all these cases, concrete examples can be found in Greece from 1821 to early 20th century, as they can be found in many other countries. A history of town planning cannot be limited to a description of the layout of particular settlements. The researcher must explore the complex mechanisms which lead to choices and explain the outcomes.

It is for these reasons that the author is convinced that the Greek 19th century experience is of interest to an international audience, particularly students of town planning in countries that have gone through similar experiences, but also researchers interested in the influence that developed countries had on nations which imported planning methods and instruments. The author relies naturally on the work of distinguished writers who have studied both sides of the spectrum: the Greek experience, but also that of countries which supplied experience and know-how. He is in addition aware that he is not a legal expert, nor historian, sociologist, economist, political scientist, surveyor, human geographer, environmentalist, or hygienist. This is an admission he regularly makes in his writings, but he still insists that town planning cannot be divorced from an interdisciplinary perspective, because it incorporates, or rubs shoulders with, these disciplines.

The mutual connections are visible when one enters the discussion of issues closely associated with spatial planning, like population movements, colonization, public administration, land management, public works, transport, or public health. Urban planning is not just about urban expansion, street alignments, and building permissions, to which Greek planning has degenerated. It is also concerned with key social, environmental, and governance problems, for which a large number of instruments have been developed. The first seeds of such instruments have been discovered in 19th century laws and planning experiments, in the form of land re-parcelling, pre-emption rights, public participation, and land use zoning. Urban planning history must uncover these precedents, but must also give room to a discussion, albeit brief, of historical events, e.g., those related to human rights, or of conditions, like that faced by the Greek state when it had to decide the disposal of the so-called national lands which it had acquired during the Greek independence war.

For the author, "turning back to the roots" was a new experience. However, since working on his original research, he had the opportunity to work further on Greek planning and to draw comparisons with the planning systems of other countries. He also had the opportunity to investigate the views of planning academics and practitioners in a more systematic way. All this enriched his approach with new visions. His thoughts in this preface are further elaborated in the introductory chapter of the proposed book.

The author would like to thank Dr. Vassiliki Angelidou who collaborated in the initial research in Greece and colleagues of the Town Planning Research Laboratory, National Technical University of Athens, for their assistance. Many thanks are due to Margaret Deignan and Ambrose Berkumans of Springer Publishers, for their help. I am particularly grateful to Prof. Frank Moulaert for his foreword.

Note: No funding was available for the writing of this book.

Athens, Greece
April 2025

Louis C. Wassenhoven

Competing Interests The author has no competing interests to declare that are relevant to the content of this manuscript.

Map of Greece

Drawn by the author

Contents

List of Figures

List of Textboxes

Chapter 1
Introduction

Abstract All chapters are divided into sections, dedicated to particular subjects. As a general rule, these sections concern international developments, domestic social and political conditions, planning activity in developed European countries, and urban planning in Greece. There are also sections with a focus on particular aspects of planning, e.g., the first planning law in Greece of 1835, the work of the Bavarian administration, the settlement of refugees, the work of famous personalities of the planning world, like Reinhard Baumeister, Josef Stübben, Camillo Sitte, Ebenezer Howard and Patrick Geddes, and British, German or French planning laws. On the whole, the structure of chapters aims at helping the reader appreciate the way town planning developments in Greece were influenced by precedents in Germany, France, England and elsewhere. All chapters start with an introductory note and end with a conclusion.

Keywords Book structure · Introduction

In writing a book on the history of urban planning in a particular country, it is of course important to study in depth planning legislation and single out all ideas, concepts and tools which gradually had a decisive effect on planning practice. Yet, one cannot disassociate the "what" the laws included and the "how" this affected town and settlement plans, from the "how" and "why" a law dictate appeared at the specific time. The factors which provide answers to these questions are in themselves quite complex. It is not sufficient to identify the person who had the responsibility to draft a particular piece of legislation or a plan. This person had certainly drawn upon his own knowledge and experience accumulated in a different country or even in the country where he was active, e.g., in Greece, but the movement of ideas is not necessarily so immediate and obvious. The protagonist of a law- or plan-making initiative could have been himself subjected to influence, whether he was a scientist or expert submitting a proposal, or the political leader who signed the final approval. These conscious or subconscious dispositions, which he may not have admitted openly, originated either in external pressures or in the tacit acceptance of political and economic realities.

L. C. Wassenhoven, *The History of Town Planning in a Dependent State*,
https://doi.org/10.1007/978-3-031-96958-4_1

Such dilemmas emerged in Greece, during the period examined in this book, for various reasons and in a variety of circumstances, frequently seemingly trivial, sometimes local, sometimes regional and national. At the local level it could be the case of tracing the street network of a city, which would require the expropriation of private land, possibly owned by persons of influence and high prestige. At the regional level, it could be the transfer of state powers to provincial authorities, the location of a new refugee settlement, or the construction of infrastructures, e.g., roads crossing agricultural estates. At the national level, the choices were related to the management of state-owned national lands, which were the only available guarantee for international borrowing, the distribution of land to war veterans of the 1821 Independence War or widows of war-victims, and the settlement of refugees from non-liberated Greek territories and of landless peasants. In such cases, the solution had to be found at a superior level of land policy, through a delicate balancing exercise.

The active application of a consistent urban planning policy, in the turbulent conditions of 19th century Greece, and its implementation on the ground, were never simple results of a clear government volition. If this was the case, voting a planning law and preparing town plans would lead to a predictable outcome and the history of town planning would be relatively easy to write. However, in practice the situation was quite different. Laws remained a dead letter or were constantly amended and plans remained unimplemented or subjected to scores of revisions. The inevitable conclusion is that planning history cannot be exhausted in a list of legal texts, of decrees published in the Government Gazette, and attractive or naive town designs. The wider context of developments and the backstage negotiations of all these policy manifestations is essential in order to comprehend planning history. To provide such a panorama of developments, ideological positions, and social mechanisms is of course a hard task. One has to widen his perspective and his range of knowledge, starting from a study of laws, plans and the existing literature on Greek history, in parallel with research on the planning models imported from foreign sources.

The influence of foreign sources depends on the form of control to which the receiving country is subjected. The country may be under outright domination of a foreign state and/or colonial power, in which case the practices it is obliged to adopt emanate from a single source. In other cases, it is under indirect control, but obliged to accept the guidance of its "protectors", because of conditions of political and economic dependence. Frequently, its own policy-makers, governors and administrators have been imbued with foreign values. The influence may be evident in multiple forms, in all government activities, such as education and social policy, but also town planning and urban development. The structure of cities can become a mirror image of social profiles divorced from traditional societal forms. The distribution of privileges can radically alter the spatial arrangements which traditional society had produced. 19th century Greece was the field of continuous interactions of such nature. It is for this reason that the author is convinced that his research would interest an international audience.

Chapter abstracts and keywords can be found at the beginning of each chapter.

Chapter 2
General Historical Overview and the Place of Town Planning and Chronology

Abstract This chapter offers a chronological guide, a kind of compass, and a concise historical outline. Here, one can find events on an international scale which set the conditions that shaped history. They range from the French Revolution, world wars and international treaties, to innovatory actions that explain developments in planning law and practice. The chronology also includes turning points in the regulation of planning and cities, like the Baron Haussmann's interventions in Paris, the conceptions of town planning of Utopian thinkers, the new towns built by Russian emperors, the design of famous open spaces and squares in Britain or Spain, the relegation of powers to German local municipalities, the British Garden Cities, and the zoning ordinances in Germany and the United States. These are just a few examples of what the reader will find in the chronological tables.

Keywords Chronology · Town planning · Historical context

2.1 Introductory Note

My aim in this chapter is to throw light on parameters which, directly or via indirect routes, affect the use of geographical space. Many present-day independent countries have reached political autonomy after centuries of subjugation to multi-ethnic empires. When they reached freedom and self-determination they had to define in their own thinking three crucial concepts: Country (or homeland), Nation, and State. Each of their citizens had a precise or vague notion of his homeland and of national identity. He also realized that he was now a citizen of an independent state, of which the rules he had to respect and obey. The territorial borders of the country had been fixed after revolutions, civil wars or the collapse of colonial empires. National identity was however a far more elusive concept, formed much earlier even when territories were enclosed within the confines of a multi-ethnic empire. The newly-born state did not necessarily coincide either with homeland, or with ethnic definitions. Country, nation and state now had to forge a single unit of existence. Greece, like many other countries, had to face this task throughout the period I deal with in this book, albeit

L. C. Wassenhoven, *The History of Town Planning in a Dependent State*,
https://doi.org/10.1007/978-3-031-96958-4_2

from my own perspective. Nation and state were not identical. Homeland, territory and government, as mental conceptions, were often clashing with nationalistic aspirations. Hellenic concepts of nationhood had roots in the fifteenth century writings of Gemistos Plethon, when the Byzantine Empire was in its death-bed. Local affairs were still administered by custom-law of Byzantine ancestry, hence the reliance on the fourteenth century legal code of Armenopoulos. The management of territory and the use of land were not handled in the style of European methods, born in the Age of Enlightenment. The task of adapting to a European model of governance was arduous, but constantly pursued by the effective rulers of the country, the Protecting Powers and by a minority of innovating Greek foreign-born *heterochthons*, in the face of resistance from the local-born *autochthons*. Some of the innovating political personalities, although of Greek extraction, did not even speak or write in Greek. Early government documents were first written in French or German, and then translated into a *sui generis* Greek idiom. I recall the observation of an academic from a former European colony that the use of the language of the colonial power had proved a real curse.

I attempt in this chapter to provide a historical framework of my analysis, without for a moment harbouring the foolish ambition of writing a history of Greece. Professional historians have done that with great success. I need however to place town planning in a historical context, without excessive details which would be tiresome for the international reading audience of planners. The chapter is also intended to prepare the reader for the chapters that will follow. Consequently, not only do I record certain events, like the arrival in Greece of the Bavarian King Otto, the ascent to power of distinguished political leaders, e.g., Charilaos Tricoupis or Eleftherios Venizelos, or legal texts of lasting impact, but I also diligently present planning innovations in other European countries. The chronological tables which constitute the main body of the chapter illustrate the periods studied in the book. Some of these periods were brief, but full of innovatory initiatives, like that of Governor Capodistria at the very beginning. Others lasted longer but were hampered by external events. International developments of a wide variety, from political alliances to town planning innovations, are an omnipresent setting of influence.

2.2 Chronology

Compiling a chronological account of the history of town planning, through the eyes of a planner, and not that of a historian, jurist or political scientist, is not a simple matter. It requires a selection of events, tendencies, laws and town planning initiatives, which cannot be exhaustive and must not deteriorate into a presentation of insignificant details. Inevitably, the choice has a personal ingredient. Prior publications by eminent researchers were of course consulted.

As stressed repeatedly, my intention was not only to present events concerning Greece, let alone to list all town plan revisions recorded in the Government Gazette, but also to produce an outline of developments on an international scale which had a direct or indirect influence. Thus, the pioneering French or German decrees on the layout of cities or the powers of local authorities were naturally included. But I also draw attention, e.g., to an international treaty which encourages economic activities with an effect on the development of cities, or to scientific advances which feed into the thinking of planners. This effect is not possible to measure. It is however beyond doubt that urban development is affected by the favourable prospect of investment attraction or that engineering, medical and social science research find its way into decisions for the planning of cities and settlements. As mentioned before, "planners" were usually foreigners or Greeks who had built their career in Germany, France, or, to a lesser extent, England, Flanders, Italy or Switzerland. In addition, not only King Otto and his Bavarian entourage, but also Count Capodistria, had accumulated experiences from Munich or Saint Petersburg. Elected Greek political leaders, like Mavrokordatos, Charilaos Tricoupis, Venizelos or Papanastasiou, also carried a load of ideas, influenced by their exposure to key countries of the European Continent. Tricoupis had at one stage represented the Greek emigrant community of London in the Greek parliament. Papanastasiou was an admirer of the English Fabian Society but had also studied in Berlin. Before coming to Greece, the multi-lingual Mavrokordatos had taught Greek to Mary Shelley!

Foreign architects and planners who worked in top positions in Greece, like Eduard Schaubert and Leo von Klenze, in the 1830s, or Ernest Hébrard and Thomas Mawson, 90 years later, carried a load of experience from Germany, France and England. Greek architects and engineers, e.g., Stamatios Voulgaris, Stamatios Kleanthis, Lysandros Kaftantzoglou, and, later, Anargyros Demetrakopoulos and Marinos Delladetsimas, who played a key role in Greek town planning, had all acquired their knowledge and experience in Germany, France and Italy.

It was not only architects and engineers whose knowledge was transferred, in a variety of ways, to Greece. It was also, e.g., physicians, Greek or not, who came to Greece, for various reasons, and had a substantial influence on the way cities were treated or planned, after being reduced to ruins and insanitary conditions during the independence war. Many of these experts found themselves near the centres of power and had certainly an influence on urban policy, even on the contents of the historic town planning decree of 1835. The exact paternity of the decree may be an open question, but there is no doubt that government policy in the period of King Otto bears the influence of physicians like Georgios Glarakis, Heinrich Treiber, or Anton von Lindermayer.

The example of physicians and hygienists brings to mind another route of influence, through foreign research and manuals, which were enthusiastically received in Western Europe, and were bound to affect the thinking of foreign physicians active in Greece in those early days. In the field of hygiene, it is most certain that persons in key positions in Greece were aware of the research of Thomas Southwood Smith, Ange Guépin, και Louis René Villermé, at the beginning of the nineteenth century, or of Franz Varrentrapp, Rudolf Virchow και Carl Reclam, in the second half of the century. When town planning is elevated to a recognisable status as a science and profession at the end of the century and the beginning of the twentieth, the teaching manuals of Reinhard Baumeister, Josef Stübben, Camillo Sitte, Patrick Geddes and

Ebenezer Howard enjoy recognition, internationally and among those able to follow European developments in Greece.

These findings guided the chronological presentation which follows. The dominant place is given to town planning, in Greece and elsewhere. Selective choices had to be made, depending on the importance of events, plans or decrees. It is certain that omissions exist, in some cases by choice. Not all the examples of industrial workers' settlements built by high-minded businessmen or planned by utopian pioneers are mentioned. Not all the plans of urban extensions in Greece are included. Most of the events mentioned in the chronological presentation are further discussed in the following chapters. I made the choice of structuring the chronology in a succession of textboxes to make it clearer.

I start with a textbox presenting international events over a long period of time, either events of global significance or narrowly related to town planning and its management.

Textbox 2.1 International events up to the outbreak of the Greek Independence War in 1821

- Fifteenth century. Rapid rise of the Ottoman Empire and occupation of Greek lands.
- 1453. Constantinople falls to Ottoman forces.
- 1480s. Redesign of the Moscow Kremlin complex, by Italian architects.
- 1489. Codification of municipal ordinances in Munich.
- 1529. Ottoman siege of Vienna.
- 1534. Royal decree in France about hygiene, street cleaning and pedestrian walkways.
- 1535. First capitulation agreement between France and the Ottomans.
- 1539. Order of king Francis I of France on the regulation of circulation and traffic problems in the suburbs of Paris.
- 1590. Creation of the royal Plaza Mayor in Madrid.
- 1605. Creation by king Henry IV of France of the Place des Vosges (formerly Place Royale), to accommodate Italian silk weavers.
- 1607. French royal decree (known as the de Sully decree), to regulate street width and impose architectural restrictions on builders.
- 1607. Decisions of Henry IV of France on street alignment, house construction, and policing of vehicle traffic.
- 1608. Foundation by Frenchmen of the city of Quebec.
- 1627. Decree drawing a border between the city of Paris and the suburbs (revised in 1672).
- 1642. New town *ex nihilo* or Ville de Richelieu in France, on a plan by the architect Jacques Lemercier.
- 1666. After the Great Fire of London, submission of a plan by the architect Sir Christopher Wren, which was not ultimately implemented.

- 1686. Foundation of a new town—extension of Erlangen—in Germany for the settlement of Huguenots.
- 1688. Plan of a new town—extension of Kassel—in Germany, by the architect Paul du Ry.
- 1680 (decade of). New fortified towns in the so-called *Zone de Fer* of France, built by Vauban, military engineer.
- 1716. Plan of Saint Petersburg by the Italo-Swiss architect Domenico Trezzini, during the reign of Peter the Great.
- 1724. Plan of the architect John Wood for Bath in England.
- 1740. French-Ottoman treaty to facilitate trade.

1765. American Revolution.

- 1767. Plan of New Edinburgh by James Craig.
- 1774. Treaty of Kuchuk-Kainarji after the Russo-Turkish war 1768–1774, guaranteeing Russian protection of Christian subjects of the Ottoman Empire.
- 1774. Industrial town Salines-de-Chaux in France, on a plan by Claude-Nicolas Ledoux.
- Eighteenth century. Town plans in several French cities (e.g., Marseille, Arras, Rennes and Limoges), sometimes accompanied by regulations of aesthetics and traffic.
- 1783. Decree on road and building construction in Paris, with regulations for stone pavements, water drainage, sewers, and street width.
- 1788. Captain Philip lands in Botany Bay, Australia, and a convict colony is founded in Sydney.

1789. French Revolution.

- 1790. Re-design of Plaza Mayor in Madrid, by the architect Juan de Villanueva, after a fire.
- 1790. Law for the founding of a Federal Capital of the United States of America.
- 1791. Thomas Paine: *Rights of Man.*
- 1791. Law reconfirming the application of the de Sully decree in France.
- 1791. Plan of Pierre Charles L'Enfant for Washington.
- 1794. Plan of the Flemish engineer Franz de Voland for Odessa, during the reign of Catherine.
- 1794. Prussian law (General Law of Prussian States) with building regulation.
- 1804. Napoleon I is crowned as Emperor.
- 1804. French Civil Law Code.
- 1806. Creation of the Kingdom of Bavaria.
- 1807. New law for the continued application of the de Sully decree in France.

- 1807. French law on servitudes due to building line recession and on recuperation of surplus values due to public works.
- 1808. Prussian law on city charters, amending the status of local government.
- 1810. French imperial decree for the protection of hygiene, through policing of dangerous, inappropriate, noxious and insanitary installations. First hint of zoning.
- 1810. French law on compulsory purchase.
- 1810. Design of Regent's Park and Regent's Street in London by John Nash and J. & D. Burton.
- 1812. Treaty of Bucharest, after the Russo-Turkish war.
- 1815. First effort in France to classify noxious activities.
- 1815. Battle of Waterloo and defeat of Napoleon I.
- 1821. Appearance of the term "Eastern Question".
- 1821. Publication of the book by Henri de Saint-Simon on the industrial system.

Before entering into the details of the presentation I summarize events of interest for the concept of Greek nationhood, which will be further explained in the text.

Textbox 2.2 The Greek distant past and events leading to the 1821 revolution

- 1345. Constantinos Armenopoulos, *Hexabiblos*—Code of Byzantine Law.
- 1348. Despotate of Morea.
- 1400$\approx$. Gemistos Plethon in Mystra—Foundations of new Hellenism. Proposal for land readjustment.
- 1415. Construction of defensive wall along the Isthmus of Corinth.
- 1453. Fall of Constantinople.
- Eighteenth century. Long period of decline of Greek lands.
- 1790s. Rigas Feraios or Velestinlis: Writings on the rise of Hellenism, published in Vienna.
- 1806. Publication in Italy by an anonymous author of the book *Greek Nomarchy, i.e. Oration on Liberty*.
- 1811. Memorandum of Count Capodistria, minister of the Russian Tzar, on the condition of the Greeks.
- 1814. Foundation of the secret Friendly Society in Odessa.
- 1820. Unsuccessful uprising against the Turks in the Danubian Principalities, under the military leadership of prince Alexandros Ypsilantis, who eventually died in Vienna.

In 1821, the war of independence against the Ottomans was declared in the monastery of Agia Lavra in the Peloponnese, raising a wave of enthusiasm among philhellenic circles in Europe and America, including the applause of Jeremy Bentham whose treatise on law was translated into Greek in 1830. The governments of the Great Powers in the European continent kept their distance, but watched the developments with attention. They were of course fully aware of the Eastern Question and of the precarious condition of the "Sick Man of Europe", i.e. Turkey. The fact remains that the first country to hail the revolution was Haiti! Soon of course, the temporary government of the revolutionaries sought the assistance of Britain, France and Russia, and a French military mission arrived in the Peloponnese. The senior English military officer Colonel Stanhope proposed in 1823 that English colonists be settled in Greece. As the revolution has its first major successes and the question of a transitional government is being debated, the name of the senior diplomat of the Russian Imperial Court Count Capodistria (Capo d' Istria) is thrown on the table of diplomatic negotiations and Capodistria visits the main European capitals in 1826.

I can now summarize important developments, mainly in Europe, concerning spatial development and town planning.

Textbox 2.3 International events and planning-related developments in Europe while Greek independence is imminent

- 1821. Simon Bolivar and independence of Venezuela.
- 1825–1830. Construction of first railways in England.
- 1820s. Urban renewal projects and construction of public buildings in Munich, frequently involving the architect of the Bavarian Royal Court Leo von Klenze, who was later commissioned to produce town plans for Athens and Piraeus.
- 1828. British Corn Laws.
- 1828. Russia declares war on Turkey.
- 1830. Revolution in Paris.
- 1830s. Massive investments in the sewer network of Paris.
- 1830. French invasion of Algeria.
- 1830. Study of Thomas Southwood Smith on infectious diseases and hygiene, soon translated into French.
- 1831. French law on municipal government.
- 1831. First railway line in France.
- 1831. International treaty on Belgium and Holland.
- 1831. Activity of the French physician Louis René Villermé, chairman of the national council of public hygiene, on the control of the Paris cholera epidemic, and the sanitary conditions of rental housing.

The textbox that follows includes the events in Greece from the outbreak of the revolution in 1821 to 1832. The main centres of the territories that were to become part of the new state were Athens, Nafplion (Nauplia), Patras, Piraeus, and Ermoupolis, on the island of Syros. With the exception of Athens, all were port cities. There were in addition eleven towns of relative importance in the Peloponnese and central Greece. Even the most important centres were in fact large villages. According to an estimate, the population of these territories declined by 29% during the years of the revolution.

The Greek uprising started on March 25th, 1821 and its main phases will be presented later, but it is of interest to start with events related to the role of the Great (or Protecting) Powers.

Textbox 2.4 International negotiations for Greek independence

- 1821. The all-powerful primate of Mani Petrobey Mavromichalis submits to Europeans a request for assistance to the Greeks.
- 1822. The revolutionaries send an official delegation to West European countries in search of a prince who would ascend the royal throne of a new Greek kingdom.
- 1825. A request is conveyed to Great Britain to act as the privileged protector of Greece, a request described by some as an Act of Subjection.
- 1825. Application to France, with the crown offered to the Duke of Nemours.
- 1825. Preliminary agreement of the Protecting Powers to act as mediators in favour of a free Greece.
- 1827. The Protecting Powers sign the first London Protocol in support of Greek independence.
- 1828. The new Governor Ioannis Capodistria arrives in Greece.
- 1828. A French military force of 14,000 men lands in Morea (Peloponnese) under the command of General Maison.
- 1829. A new London Protocol defines the borders of the new state and the subsequent Treaty of Adrianople, co-signed by the Ottomans, consolidates Greek independence.
- 1830. A third London Protocol confirms Greek independence and the state's frontiers.
- 1831. General Schneider of the French military mission submits proposals for the creation of a French colony in Western Peloponnese and for the settlement of French citizens, who would import the necessary capital to be invested in productive activities.

We can now turn to domestic developments and government actions and decrees, during the same period between 1821 and the confirmation of the status of the new state.

Textbox 2.5 Domestic developments during the independence war

1821

- Assembly of the kodjabashis (primates and notables) and constitution of the senate of Messenia.
- Assembly of all the kodjabashis of Morea (Peloponnese) and constitution of the senate of the Peloponnese.
- Arrival in Greece of Demetrios Ypsilantis, brother of Alexandros, and of Alexandros Mavrokordatos, who were destined to play a key-role in Greek affairs.
- Local regimes are put in place, i.e., the senates of west and east continental Greece and of the Peloponnese.
- Tripolis (Tripolitsa)—Ottoman capital of the Peloponnese—is ransacked by the Greek rebel forces under the leadership o Theodoros Kolokotronis, and the Turkish population is slaughtered.

1822

- 1st National Assembly of Epidaurus (near Corinth), first Greek Constitution, and first national Executive Government.
- Massacre of the people of the island of Chios, which inspired a poem by Victor Hugo and a painting by Delacroix.

1823

- 2nd National Assembly of Astros and election of a deputation (parliament).

1824

- Civil wars among the Greeks, while the revolution is still in progress.
- Devastation of the islands of Psara and Kassos.

1825

- Organization of a system of tribunals.
- Street layout plan of town of Kalamata by French volunteers.
- The new parliament decides the sale of part of national land occupied by the revolutionary forces.

1826

- Siege by the Turks of the town of Messolonghi and exodus of the besieged population, with Lord Byron sharing their ordeal.
- 3rd National Assembly at Epidaurus, retraction of the decision to sell national lands, and formation of a new Administration Committee, as temporary government.
- Siege of Athens by the Turks.

1827

- Death in battle of Georgios Karaiskakis, General of the Greek forces in the battle of Athens.
- Occupation of the Acropolis of Athens by the Turks.
- Special National Assembly at Troezen, new Constitution, and invitation to Ioannis Capodistria to assume the duties of Governor.
- Naval battle of Navarino, where the allied fleets of the Protecting Powers destroyed the Ottoman fleet.
- The national assembly continues its proceedings in Nafplion, which becomes temporarily the seat of a counter-government.
- Aegina is designated as the state capital.
- The Troezen assembly invites Greek expatriates to form colonies in Greece and orders a public auction of perishable national properties.
- Land distribution to war veterans and, later, to widows and orphans of fighters killed in battle.
- Decrees are issued for expatriate settlements and the founding of new towns, including one near the Corinth Isthmus.

1828

- Count Capodistria takes over the duties of Governor.
- Establishment of "Panhellenion", a version of parliament, and of a council of ministers.
- New administrative structure: Prefectures, provinces and municipalities.
- Decree on new courts of justice.
- Decree on the application in Greece of the French commercial code.
- Creation of a National Monetary Bank.
- The German physician Treiber, an ardent philhellene, is appointed as director of the Nafplion military hospital.
- Town plan of Tripolis, drawn by Stamatis Voulgaris, a Greek engineer and officer in the French army, considered as the first Greek town planner, and the Frenchman Garnot, military engineer.
- Voulgaris is commissioned to design the plan of Pronoia, a suburb of Nafplion.
- Proposal of Trieste expatriates to found a new town on the coast of Glarentza in western Peloponnese,
- In 1828–1829, plans for Patras, Nafplion, Corinth, Messolonghi, Argos and Pylos, by Voulgaris and various foreign engineers (Audoy, Peytier, de Schaumbourg, de Boroczyn).

1829

- The capital of the new state is transferred to Nafplion.
- 4th National Assembly at Argos.

- Administrative division of the country into 13 departments.
- Decrees for the organization of the Senate and the "ministerial system".
- The Governor orders a survey of national lands.
- New penal code, based on French model.
- Creation of an experimental farm in Tiryns, near Nafplion.
- The engineer Vallianos is appointed as architect, in charge for the implementation of the Nafplion plan.
- Foundation by the American physician S.G. Howe of the short-lived colony Washingtonia near Corinth.
- Decision of the Governor setting up departments of fortification engineering and of architecture.
- Submission by Voulgaris of the plan for Patras, the first comprehensive town and street layout plan.
- Decree on the renovation or demolition of dangerous buildings.
- Official approval of the town plans of Methoni,near Kalamata, by Audoy, Corinth by Peytier and Tripolis by Voulgaris and Garnot.

1830

- Decree recognizing the equality of custom-law and formal laws and limiting the application of French law. Reference to the use of Byzantine law.
- Governor's guidance on the importance of private property and the protection of small property owners.
- Governor's order prohibiting the exportation of antiquities.
- Decree authorizing the sale of perishable public properties.
- Appointment of the Greek architect Stamatios Kleanthis and the German architect Eduard Schaubert, both educated in Germany, as Government Architects.
- Decrees on wills and testaments, custom-law validity, and horizontal land property, implying the application of surface rights.
- Government orders for the implementation of the town plan of Patras, sanctioning the exchange of land properties in a form of land readjustment, revoked later in the 1850s.
- Authorization of a French army colonel to execute a rehabilitation project of the town of Methoni, a military stronghold near Kalamata.
- Decree for the foundation of the new town of Krissa, near Itea in Central Greece, on a plan by A. Kalandros.
- Governor's order to encourage those wishing to erect houses in the town of Argos.
- Governor's order for the implementation of the plan of Patras.
- Approval of the plan by R.D. de Boroczyn for the town of Argos.
- Decree to secure the resources necessary for public use spaces in towns, such as streets or water reservoirs.
- Order on the granting of building permissions.

- Governor's guidance on the compensation of the residents of the town of Galaxeidi, affected by street construction.

1831

- New plan for the town of Pylos.
- Decrees on the settlement of Cretan refugees, persecuted by the Turks, and on the erection of new towns and suburbs, through concession of land to those wishing to build.
- Instruction of the Governor to the Senate for the settlement of refugees from Crete.
- Decree on the building of the Pronoia suburb in Nafplion.
- Assassination of Governor Capodistria by members of the Mavromichalis family, the Mani's clan of *kodjabashis*.

The internal feuds between rival factions continued during the governorship of Count Capodistria, leading to his tragic death. The rivalries were made clear in civil wars and the antagonistic National Assemblies, convened by opposing parties. The time was ripe for the Protecting Powers to place a nobleman of their choice on the royal throne of Greece, preferably from a family other than their own royal families. I will list the events of the period of the kingship of the Bavarian Prince Otto, which follows an interim period under a temporary governor. However, as I did before, I will first present international developments, which serve as a useful background. Included here are also actions which show the extent of external domination in Greece or treaties with an impact on internal affairs.

Textbox 2.6 International developments while Greece was under Bavarian administration

- 1832. Constantinople Treaty grants to Turks leaving Greece the right to sell their properties in Attica to private individuals, a golden opportunity for investors.
- 1832. Thousands of victims of a cholera epidemic in Paris.
- 1832. Research of the French professor of medicine Ange Guépin, on the cholera epidemic in Nantes.
- 1833. English law on child employment.
- 1834. Publication in Greek of the treatise of Jeremy Bentham on law.
- 1834. Revolt of silk weavers in Lyons.
- 1835. Building regulations in Düsseldorf on the height of buildings, later revised in 1855.
- 1835. Law on Municipal Corporations in England.
- 1836. Foundation of Adelaide in South Australia, with a centre designed by William Light on a gridiron pattern.

- 1839. Beginning of *Tanzimat* (reorganization) social reforms in the Ottoman Empire, which included, among others, equality for all religious communities.
- 1840. Work by P.J. Proudhon on the nature of property.
- 1841. Decree on the prohibition of construction in a peripheral zone of Paris and French law on land expropriation for public benefit.
- 1841. Statement of the British ambassador that a fully independent Greece is an absurdity.
- 1842. *Report on the Sanitary Condition of the Labouring Population* in England, by Edwin Chadwick.
- 1842. Devastating fire in the centre of Hamburg.
- 1842. Plan by William Lindley on the Hamburg drainage and water supply infrastructure network.
- 1842. Treaties of Nantes and opening of Treaty Ports in China for foreign colonists.
- 1844. Uprising of German textile industry workers.
- 1845. Book by Friedrich Engels on the condition of the working class in England.
- 1847. English laws on urban improvements, policing, water and gas infrastructures, cemeteries, markets and exhibitions.
- 1848. Revolution in Paris, dethronement of King Louis-Philippe, second French Republic, and dismissal of Metternich in Austria.
- 1848. Communist Manifesto by Marx and Engels.
- 1848. French law prohibiting the rental of insanitary houses.
- 1848. English public health law, followed, in the nineteenth century, by a series of laws on working class housing.
- 1849. Liberal constitution in Denmark.
- 1850. Economic reforms in Piemonte by Camillo Cavour.
- 1850. The "Don Pacifico Affair", concerning an English citizen, whose house was burned in Athens during a riot, and provoked a British intervention and Palmerston's phrase about the "long arm of the Empire".
- 1850. French law on insanitary housing, considered by some historians as the first French town planning law.
- 1851. Saltaire, industrial workers' town built in Yorkshire by Titus Salt, owner of a wool textile manufacturing plant.
- 1852. A French legislative decree on the road network of Paris, compulsory land acquisition and building regulations, gives the prefect Baron Haussmann the legal means to implement his renovation plans in Paris.
- 1853–1870. Reconstruction of the Paris centre and extensive building demolitions.
- 1853. New building code in Berlin and new system of water supply and street cleaning.

- 1854. A British fleet under the command of captain Parker blockades the port of Piraeus (The "Parkerika" affair).
- 1854. Vienna treaty on claims over previous Ottoman territories in Eastern Europe.
- 1856. Paris treaty on the integrity of Turkey.
- 1855. Prussian ministerial decision on the design and construction of city buildings and on the right of municipalities to fix building lines.
- 1856 *Hatt-I Humayun* (Sultan's ordinance) and 1858 Ottoman law on land changed the status of rural workers cultivating the soil of *chiflik* estates.
- 1857. Innovations are introduced in the planning of town extensions of Vienna, followed by a town planning competition of the peripheral zone of the city walls, leading to the opening of a ring road (*Ringstrasse*).
- 1858. Year of the Great Stink in Central London, caused by human waste and industrial effluents.
- 1858–1862. Plan of James Hobrecht for Berlin and construction of new street network.
- 1859. Plan of Ildefonso Cerdà for the extension of Barcelona.
- 1859. Industrial workers' settlement, known as *Familistère*, built by the industrialist Jean-Baptiste André Godin in Guise of France.
- 1860 and after. Appearance in Germany of housing complexes known as Barracks for Rent (*Mietskaserne*).
- 1861. Proclamation of Kingdom of Italy.
- 1864. Workers' housing settlement built by the industrialist Jean Dollfuss in Mulhouse of France.

The Bavarian period, following the chaotic interval after the assassination of Ioannis Capodistria, is full of examples of innovations and external pressures. It is also marked by the co-existence in Athens of people from various regions, speaking local dialects incomprehensible to their neighbours. In 1836 a theatrical play by D. Vyzantios was performed, called *Babylonia*, making fun of this communication problem.

Textbox 2.7 The years of the Bavarian Regency and King Otto up to 1844

1832

- 5th National Assembly at Argos and election of Avgoustinos Capodistria (brother of Ioannis) as president of an interim government.
- An international loan is contracted under the guarantee of the Protecting Powers.
- Parallel assembly in the Pronoia suburb of Nafplion and useless decree on the issue of national lands.

- Formation of a public finance bureau, as annex of the Ministry of the Interior, headed by the Frenchmen Roujoux and d'Eichtal.
- 1832–1835. Period of government by the Regency triumvirate of Count Josef Ludwig von Armansperg, Georg Ludwig von Maurer and Carl August von Abel.
- Town plan for Athens by the architects Stamatis Kleanthis and Eduard Schaubert.

1833

- King Otto arrives in Greece.
- New secretariats (ministries) are created.
- Introduction of a formal Government Gazette (GG).
- Royal Decree establishing Athens as the seat of government.
- A public finance office is created, headed by a French economist.
- Introduction of 3-member courts of justice in three provincial towns.
- 1833–1835. Approvals of penal law and civil and criminal law procedures, modelled on French and German prototypes.
- Royal decree on the reconstruction of the city of Athens and the transfer of the capital.
- Decision on the opening of new roads in various parts of the country.
- Royal decree on penalties related to street cleaning and building maintenance.
- Royal decree on the powers of prefects, the administrative structure of prefectures, provinces and municipalities, modelled on the French precedent, and the creation of municipalities, which remained in force until 1912.

1834

- Royal decree confirming the creation of the Public Finance Bureau.
- Royal decree establishing Athens as the seat of the Royal Court.
- Armed insurrection in Mani.
- Controversial trial of the popular revolutionary leaders Kolokotronis and Plapoutas.
- Formal law on the new municipalities.
- The Regency member von Maurer, in charge of justice, is revoked back to Bavaria, by order of King Ludwig of Bavaria.
- The town plan of Nafplion is approved by royal decree.
- Bureaus of architecture and surveying are created headed, respectively, by Eduard Schaubert and D.A. Guebhard.
- Royal decree on the distribution of land to military personnel and war widows.
- Decision on the settlement of Bavarians, which is implemented only in one location in Attica.

- Supplementary decrees about the plan of Athens.
- A new plan of Athens is designed by Leo von Klenze, architect of the royal court of Bavaria.
- Royal decree for the creation of the town of Sparta, satisfying the wish of the King of Bavaria.
- Submission of a proposal by Baron von Rheineck for the settlement of German colonists.
- 1834–1843. Long series of plans for various towns and new settlements.

1835

- The French commercial code appears in a Greek translation.
- Royal decrees on
  - Civil law, approved temporarily until its revision.
  - Mortgages.
  - Regulation of private building in the port of Piraeus; practically approval of a new town.
  - Family dotation out of national land properties.
  - Concession of land plots in Piraeus to refugees from the island of Chios.
  - Plan of the old quarters of Athens.

Royal decree regulating the sanitary construction of towns, townships and villages:
Turning point in the history of Greek planning law.

- Plan of Piraeus by the architects Kleanthis and Schaubert.
- Law on land grants to families and municipalities and on the provision of public spaces in towns.
- Demolition of the so-called "Ali Hasheki" fortified Turkish wall around Athens.

1836

- Royal decrees on
  - Christianized Muslims and their properties.
  - Ecclesiastical properties.
  - Taxation of state-granted privileges.
  - Provincial councils.
  - Municipal police.
  - Control of infectious diseases.
  - Creation of a university in Athens.
- Law on the establishment of a National Bank.
- Royal decrees on
  - Implementation of the plan of Athens.

- Settlement of Chios refugees in Piraeus.
- Settlement of refugees from Psara in Eretria.
- Taxation of buildings.
- Sale of national lands and perishable properties.
- Private forests.
- Cadastral survey.
- Public buildings in government use.
- Architectural education.
- Compensation for land acquired for the opening of streets.

1837

- Drought and water shortage in Nafplion and building of aqueducts.
- Establishment of a School of the Arts, with the Bavarian military engineer Friedrich von Zentner as director.
- Approval of the plan of the port-town of Ermoupolis on the island of Syros.
- Topographical map of Athens compiled by the German officer Aldenhoven.
- Royal decrees on
  - Settlement on the small island of Elafonissos, off the coast of Laconia in southern Peloponnese.
  - Settlement of Chios refugees in Piraeus.
  - Amendments of the Athens plan.
- Royal order for a settlement of Bavarians in Stefani, near Athens, which was never implemented.

1838

- Law on land grants to members of the Phalanx, a war-veterans' association.
- Royal decrees on
  - Plan of Ermoupolis, which adopted building regulations applied in Athens, and extension of the same regulations to the entire national territory.
  - Settlement in Piraeus of migrants from the island of Hydra.

1839

- Proposal by the architect Lysandros Kaftantzoglou to open an avenue linking Athens with Piraeus and the sea-front, thus orienting urban expansion towards the coast.

1840

- Construction of the first sewer in Athens, in the vicinity of the royal palace, allowing the diversion of effluents to an existing water stream

1841

- Report, 180 pages-long, by D.A. Guebhard head of the land surveyors' bureau.

1842

- Royal decrees on
  - Amendments and implementation of the Athens layout plan.
  - Adoption of the Athens building regulations in all towns, townships and villages of the kingdom, a decree amounting to an early quasi-national building regulation code.
  - Ermoupolis plan, setting minimum standards for plot faces on access streets, plot acreage and other details.

1843

- Legislation for the protection of land properties of peasants.

Military coup against King Otto:

Bavarians are removed from government posts, a national constitutional assembly is convened and a new Constitution is voted.

- Decree about the School of Arts, of which Kaftantzoglou is appointed as director after the German von Zentner is suspended from duties.

The following 20 years, i.e. until King Otto is evicted and the Danish William (Vilhelm) of Schleswig–Holstein-Sonderburg-Glücksburg is crowned as George I, were bound to be unstable. The democratic movement was dissatisfied with the limited reforms of the 1843 Constitution. Routine government activities continue in all fields, town planning included, but the period is clearly one of transition, while political instability grows and the state finances deteriorate. National dreams for the liberation of Greek lands still under Turkish occupation are fuelled.

Textbox 2.8 The days of constitutional monarchy under King Otto

1844

- Ioannis Kolettis, prime minister and advocate of a "French connection", coins the term "Great Idea", symbolizing the irredentist dream of the nation.

1845

- Proposal of the Saint-Simonist d'Eichtal for the drainage of Lake Copais and its conversion to cultivable land.

1847

- Law on municipal taxes introduces the right of port towns to impose gate tariffs.
- Royal decree on the settlement of refugees from the island of Psara in Eretria, on the island of Euboea.

1848

- Law on three settlements to be colonized by refugees from Crete.

1850

- Cholera epidemic in Athens, Piraeus and Ermoupolis, the most important urban centres of the country.
- Constant stream of decrees amending the plans of Athens and Piraeus, even in the "garden of the people" (formerly royal garden) of Athens.

1852

- Royal decrees on
  - Demolition of buildings for violating the building line and of dangerous structures.
  - Police duties in Athens and Piraeus for issuing building permissions.
  - Amendment of the plan of the port city of Patras, with the addition of rows of building blocks.

1855

- Law on illicit construction on illegally occupied land and payment of fines.

1856

- Civil law regulating matters of land property and use, land possession, freehold property, usufruct, testaments, transaction deeds, and residential properties.
- Law on pavements, sewers and open land belts around cities.
- Royal decree on the urban belt around Athens.

1857–1859

- Report on problems of public economic and fiscal management and tax collection by the Allied Joint Financial Commission, appointed by international creditors.

1857

- A French mission is invited by the government to produce a geodesical study of Athens.

- New extension of the Athens plan, north of the palace, allegedly only for agricultural use.

1858

- According to data published in the Government Gazette the population of the country has surpassed the 1 m. threshold.
- Law on the new town of Corinth, following a disastrous earthquake.
- Royal decree on pedestrian ways.
- Law on real estate properties in the city of Patras, fixing procedures of compensation for compulsorily acquired land plots, in lieu of land exchange or maintenance of surface rights.
- First bridge over the Evripos straight, linking mainland Greece with the island of Euboea.
- New town plan for the town of Lamia.

1859

- Law on the concession of national land plots in towns, for the creation of food markets.

1860

- Plan of the Stavridis Commission for the revision of the von Klenze plan of Athens.
- Royal decree for a new coastal settlement, near the town of Kalamata.

1860s: Political instability and dominance of clientelism.

1861

- Re-confirmation of the 1852 decree for the extension of the plan of Patras.
- Royal decree for a new town in Kastri, in Central Greece, near a mineral site.

1862

- Municipal elections, with voting rights for all males.

After a new revolt, King Otto is deposed.

1863

Plebiscite in Greece and ascent to the throne of King George I.

- Reforms in the School of Arts, under a new Greek director educated in France.
- Royal decree about a city belt around the city of Piraeus.
- Statute regulating the shipyards of Ermoupolis.
- Statute regulating the building blocks around the Athens Omonoia (Concord) Square, originally planned as a location of the royal palace.

The fluidity of the balance of power in Europe continued while Greece was adjusting to new political conditions, but the clear trend was a movement towards an increasing British influence in Greek politics. Dependence on Britain was rising as the power of the British colonial empire was growing, while Greece was moving towards an economic crisis and military misfortunes. Frictions were also due to incidents like the murders by bandits in 1870, in Dilesi, of English travellers and diplomats. Such events did not prevent British companies from investing in Greek projects or the British government making the gesture of ceding the Ionian Islands to Greece, as we shall see later. However, the international setting is also full of political developments in Germany, France or the Balkans, and town planning innovations and initiatives. This is the time when town planning enters a phase of consolidation as a science and profession.

Textbox 2.9 International developments during the reign of George I in Greece, until the end of the nineteenth century

- 1865. Construction of Ringstrasse, around the old city of Vienna.
- 1866. German Confederation under Prussia.
- 1868 (and 1875). English laws for the improvement of the housing conditions of working classes and public health (Torrens' Act and Cross's Act).
- 1870. Several German *länder* transfer the responsibility of road building and street alignment to local authorities.
- 1871. Constitution of the German Empire.
- 1871. Rising of the Commune in Paris.
- 1873. Conference in Frankfurt of the *Deutscher Verein*, partly dedicated to hygiene.
- 1874. In Cologne, guidelines are issued for architects and engineers on city extensions.
- 1874–1875. Prussian laws on municipal powers (expropriations, street-making, street frontage of buildings).
- 1875. English Public Health Act and housing act, authorizing local authorities to expropriate land. On the basis of the health act, 700 health agencies were created in cities.
- 1876. Publication of the first town planning manual by Reinhard Baumeister.
- 1877. British colonial power at its apex: Queen Victoria is proclaimed Empress of India.
- 1878. Treaty of Berlin: Turkish loss of territories in the Balkans.
- 1878. Zoning is introduced in the plan of Dresden, with the designation of three zones where industrial use is prohibited.
- 1879. The city of Cologne is empowered to manage issues of streets and traffic.
- 1880. Town planning competition for Neustadt of Cologne.

- 1880. Publication of the book *Progress and Poverty* by Henry George, in which he argued that land property should the exclusive basis of taxation.
- 1880s. Work of Charles James Booth on the miserable living and working conditions in London.
- 1881. Plan of Josef Stübben for the Neustadt of Cologne.
- 1882. First emergence of the Triple Alliance between Germany, Austria and Italy.
- 1883. French law on the need of research before a project is approved for reasons of public benefit.
- 1884. Municipal law in France on the need of street layout plans in all communities.
- 1887. French law on the preservation of historic buildings, which was revised in 1913.
- 1888. Industrial workers' settlement Port Sunlight, built by the soap manufacturer William Hesketh Lever in Merseyside.
- 1888. Establishment in England of County Boroughs, independent of the old County Councils.
- 1889. Book on town planning by Camillo Sitte.
- 1890. Book by Josef Stübben (*Der Städtebau*).
- 1891. Franz Adickes is elected Mayor of Frankfurt, a post he held until 1912.
- 1891. Full zoning imposed in the plan of Frankfurt.
- 1891. Electricity-driven public transport in the German town of Halle.
- End of nineteenth century: Population explosion in London, Paris, Berlin and Vienna.
- 1892. Cholera epidemic in Hamburg.
- 1892. Town planning competition for the plan of Munich and proposal by Karl Henrici.
- 1892. Building regulation of the Prussian government for Berlin and severe zoning restrictions.
- 1892. Land readjustment implemented in Hamburg.
- 1893. Law of the Prussian parliament on land readjustment, known as *Lex Adickes*, implemented only in Frankfurt.
- 1893. New building regulation for Cologne, imposing four population density zones.
- 1894. Law Jules Siegfried in France imposes on the state the obligation to deal with the problem of social housing.
- 1894. The London Building Act introduces the public control of the opening or widening of streets, railway lines construction, open spaces around buildings and building height.
- 1896. Law on local streets in Baden.
- 1897. Patrick Geddes works in Cyprus during the settlement of Armenian refugees.

- 1897. German law on cadastral survey.
- 1898. French law on safety in cities and policing of derelict buildings.
- 1898. Book by Ebenezer Howard on garden cities.
- 1899. Creation of the Garden City Association in England.
- End of nineteenth century. The term *urbanisme* appears in France.

In Greece itself the impact of international developments is clearly evident in political terms, although not so visible in the field of town planning, except in the cases of towns in newly annexed territories and amendments to accommodate the execution of public works. Relevant adjustments take time and are delayed by the slow pace of adaptation to a changing geopolitical environment. There is however ample evidence of social change, economic modernization, and of a rise of new productive forces. Only selected representative examples of town planning interventions are mentioned.

Textbox 2.10 Greece: The period of the reign of George I to the end of the century

1864

- New Constitution, upgrading of the role of local mayors and reference to the distribution of national lands.
- Annexation of the Ionian Islands.
- Mineral extraction rights in Lavrion (site of ancient silver mines near Athens) are acquired by the Roux-Serpieri Company.

1865

- Law on dried raisin production due to gain in importance, as influential producers press for its expansion to agricultural land distributed to landless peasants, hence leading to the fragmentation of land properties.
- Successive decrees amending sections of the plan of Athens.
- Law on the erection of buildings on national and church-owned land.

1866

- Concession of state-owned land to the shipyards of Ermoupolis.
- Repeated decrees on the plan of Nafplion and the demolition of the surrounding defence fortifications.
- Law on new settlements and the necessary town plans.
- 1866–1869. Cretan revolt against the Turks.

1867

- Contract with foreign firm and law sanctioning the agreement for the drainage of Lake Copais.
- New Royal decree on the street layout plan of Nafplion and the demolition of the city walls.
- Royal decree on dangerous decrepit buildings.
- Law on "the implementation of town and township plans of the Greek Kingdom".
- First approval of the plan for an industrial workers' settlement in Lavrion.

1868

- Law on road construction and classification of roads into national, provincial, municipal, and rural.

1869

- Law on the digging of the Corinth Canal, of major importance for maritime traffic.
- Southward extension of the Athens plan to the district of Patissia.
- Decree determining building regulations in the town of Argos.

1871

Law on the distribution of national lands and accompanying executive order brings to an end the decades' old government indecision.
Law on the settlement of New Corinth, many years after the 1858 earthquake.
Creation of the Music and Dance Conservatoire of Athens.

1873

- New decree on the demolition of the Nafplion walls.
- Law on the new Corinth settlement and applications for land grants.
- Stock exchange collapse of the shares of the Lavrion mineral extraction company and subsequent Royal decree.
- Royal decree for the "catharsis" of the royal summer palace of Corfu!

1875

- Introduction of the obligation of governments to secure a confidence vote in Parliament (principle of "declared" majority).
- Prime Minister Charilaos Tricoupis ascends to power for a first term.
- Map of the Athens basin and Attica by Johann August Kaupert.
- Time-extension of the decrees on the settlement of refugees from the island of Psara, many decades after the independence war.
- Inauguration of the Aedipsos Spa, followed by the Ypati Spa in 1896 and the Kamena Vourla Spa in 1897.

1880

- Abolition of tithe taxation imposed on farmers.
- 1880–1893. Digging of the Corinth Canal, which favours the port of Piraeus, at the expense of the port of Ermoupolis.

1881

- Annexation of Thessaly, a region of great importance because of its agricultural plains.
- Approval of contracts with foreign companies for railway construction.

1882

- An advisory mission of French engineers is invited to assist the Greek government.
- Establishment of the Public Health Society.
- Establishment of a share company for public and private works.
- Laws on the drainage of lakes, particularly for that of Copais.
- Law on mine ownership and mineral extraction rights.
- A contract for the construction of the Piraeus-Patras railway line is sanctioned by law.
- Establishment of the share company "Piraeus, Athens and Peloponnese Railways".
- The contract for the construction of the Lavrion-Athens railway line is approved by law.
- Decree on the construction of cesspools.
- Law on the appointment of surveyors (geometers) in pubic administration.
- Royal decrees on
  - Pavements and sewers in the city of Volos in the region of Thessaly.
  - Street layout of Kifissia, fashionable suburb of Athens.
  - Amendments of the Athens plan, particularly around Lycabettus hill.
- Law on land grants to Greek citizens in approved settlements.
- During 1882–1885, plans are produced for the cities of Volos, Larisa, Karditsa and Trikala, in the newly annexed region of Thessaly.

1883

- In 1883 and 1887, the first tobacco manufacturing plants open in Athens.

1884

- Decree on building regulations accompanying the town plans of Athens and Argos.

1886

- New electoral victory of Charilaos Tricoupis.
- Decree on government bonds.
- Amendment of the Athens plan to accommodate the Athens-Lavrion railway line.
- Repeated amendments (1886–1893) of the town plans of cities in Thessaly.

1887

- Law on cigarette manufacturing.
- Laws on customs levies and on professional activity and building construction taxation.
- Contract with Lake Copais Co. Ltd, for the drainage of Copais Lake.
- Decree on deadlines for the completion of digging works of the Corinth Canal.
- Law on the corps of Civil Engineers.
- New adaptations of the Athens street-layout plan, to provide access to new railway stations.
- Royal decree on pavements and sewers of the town of Karditsa in Thessaly.

1888

- Law on road building and related projects.
- Law on compulsory land purchase and compensations for road projects.

1889

- Amendment of the town plan of Argos.

1890

- Fire in the Ottoman-occupied city of Thessaloniki.
- Auction for the new Evripos bridge project.

1892

- Replanning of the burnt section of Thessaloniki.
- Inauguration of the drainage project of Lake Copais and industrial settlement in nearby Aliartos.
- Severe international crisis of raisin trade, with serious impact on Greek economy, especially in the Peloponnese.

1893

- Law on commercial and manufacturing trademarks.
- Brewery plant of Carolos Fix on Syngros avenue in Athens.
- Inauguration of the Corinth Canal project.

Financial bankruptcy and famous announcement by Prime Minister Tricoupis: Unfortunately, we are bankrupt!

- The Athens plan is amended for the umpteenth time.
- Amendment of the plan of the Athens coastal suburb of Palaion Faliron.
- Decree on pavements and sewers of the town of Pyrgos (Western Peloponnese).

1895

- Prime minister Theodoros Deliggiannis.

1896

- The first modern Olympic Games are held in Athens.
- Inauguration of Evripos revolving bridge.

1897

- War with Turkey and military defeat. Payment of compensation to Turkey and supervision of public finances by an International Control Commission.

1898

- Declaration of an Autonomous Cretan State, under international supervision.

1899

- Constitution of the Cretan State.

The period from the end of the nineteenth century to the Greek Asia Minor Disaster of 1922 is full of bloody wars, international treaties and negotiations, revolutions, path-breaking town planning legislation, and initiatives for the consolidation of the planning science and profession.

Textbox 2.11 From 1900 to the Greek-Turkish Asia Minor war: International review of events

- 1899. Germany secures Baghdad railway contract.
- 1900. Building regulation in Saxony with clauses on zoning.
- 1900. German Civil Code and law on urban design.
- At the beginning of the century:

– Town plans for Krefeld, Mülheim an der Ruhr, and Düsseldorf, work of the landscape architect Adolph von Vagedes.
– Influence of the book by Josef Stübben on Russian architect-planners,
– Garden city experiments in Germany.

- 1902. Renewal of German, Austrian and Italian Triple Alliance.
- 1902. French law on public health lays the foundations of a system of policing sanitary conditions, municipal and prefectoral health regulations, compulsory for builders.
- 1902. French edition of the book by Camillo Sitte.
- 1903. Creation of the French association of *Cités Jardins.*
- 1903. Letchworth Garden City on a plan by Barry Parker and Raymond Unwin.
- 1904. Systematic land policy in Frankfurt to combat land speculation.
- 1904. Report by T.C. Horsfall, an English observer, praising the German town planning system.
- 1904. Goerke and Sitte provide a definition of town planning.
- 1904. Anglo-French *Entente Cordiale.*
- 1906. Hampstead Garden Suburb in London, on a design by Raymond Unwin.
- 1906. Frankfurt, Ulm and Freiburg municipalities follow a policy of purchasing land,
- 1906. French law, amended later in 1930, for urban and natural landscape protection.
- 1906. First issue of the first town planning journal (*Städtebau*).
- 1906. The English term "town planning" appears for the first time.
- 1907. New edition of the book by Josef Stübben.
- 1907. Lethal famine in Russia.
- 1908. Revolution of the Young Turks against the Sultan.
- 1909. First Town Planning Act in Great Britain, revised several times later, with the 1932 act changing the name to Town and Country Planning.
- 1909. Book by the American Benjamin C. Marsh, entitled *An Introduction to City Planning.*
- 1910. First British journal on town planning, under the title *Town Planning Review.*
- 1911. The French society of *urbanistes* is founded, with members of the Social Museum playing a key-role, including Ernest Hébrard, who later worked in Greece.

1912–1913. Balkan Wars against Turkey.

- 1912. Public offices for the promotion of cheap housing are founded in France (*Loi Sellier*).
- 1912. Italy occupies the Dodecanese Islands in the Aegean Sea.

1914. The assassination of Archduke Ferdinand of Austria in Sarajevo becomes the first episode of World War I, the "Great War".

A succession of declarations of war follows.

- 1914. The Town Planning Institute is founded in England, much later renamed Royal Town Planning Institute.
- 1915. Publication of the book *Cities in Evolution*, by Patrick Geddes.
- 1916. Zoning ordinance of New York.
- 1916. First town planning manual in French, by the Belgian L. Van der Swaelmen.
- 1916. The *Entente Cordiale* allies occupy Athens.
- 1917. Greece enters the World War.
- 1917. Two town planning treatises in French, by Auburtin & Blanchard and by D. A. Agache.
- 1917. Foundation of the American City Planning Institute.
- 1917. French law on dangerous and insanitary activities, revised in 1932.
- 1918. First global planning law in Prussia on housing, which included elements of urban design and planning.
- 1918. French *Loi Chauveau*, which recognized the principle of public benefit resulting from land readjustment (*remembrement*).
- 1918. Ceasefire of Moudros, and termination of the Great War—although not for Greece!
- 1919. Mussolini founds fascist movement in Italy.
- 1919. German regulation for long-term property leases.
- 1919. French *Loi Cornudet*, the first real town planning law in France, according to which all towns above 10,000 inhabitants had to produce a PAEE plan (*Projet d' Amènagement, d'Embellissement et d' Extension*).

1919. The Treaty of Versailles seals the end of the Great War.

- 1920. Victory of the Bolsheviks in Russia.
- 1920. Welwyn Garden City in England, on a plan by Louis de Soissons.
- 1920. The League of Nations is founded.
- 1920. Treaty of Sevres, recognizing the territories of Macedonia, Epirus, Thrace and Izmir region as being Greek soil. The Greek Great Idea of the 2 continents and 7 seas if briefly realized.
- 1923. The Treaty of Lausanne, after the defeat of the Greeks in Asia Minor, restricts Greek territorial gains
- 1922–1924. Modern town planning movement and proposals of Le Corbusier and Hilberseimer.

Some of the events listed in the previous textbox will reappear in the next one, in which domestic developments in Greece are included. If it was not for the reconstruction of Thessaloniki after the fire of 1917 and the most-important planning Legislative Decree of 1923, developments in the town planning field would be insignificant. The year 1923 is of course the beginning of the huge effort to provide settlements for the gigantic inflow of about 1.3 million refugees who migrated to Greece after the Asia Minor Disaster.

Textbox 2.12 Greece after the turn of the century until the arrival of refugees from Asia Minor

- Turn of the century: Important manufacturing industries are beginning to appear in the production sectors of cement, fertilizers, chemical paints, drugs and engineering machines.

1901

- Decree of the independent Cretan State on town plans for towns, townships and villages.
- Town plans for the towns of Chania and Herakleion are published in the Official Gazette of the Cretan State.
- Laws of the Cretan State on compulsory land acquisition, public works, auctions and project implementation.

1905

- Assassination in Athens of the former Prime Minister Th. Deligiannis.
- Cloud of dust over Athens.
- Cretan insurrection in Therissos.

1906

- Beginning of massive emigration to America.

1907

- Foundation of the Association of Greek Manufacturers and Craftsmen.
- Law on settlements and land distribution in Thessaly and creation of the Thessaly Agricultural Fund.

1908

- Royal decree on the street layout of Athens and division into neigbourhoods.

1909

- Coup of the Military League in the camp of Goudi, near Athens, against the royal regime.
- Revisionary constitutional assembly.
- Law on road construction and road classification.
- Law on administrative division of the state.
- Town planning law, with special emphasis on the plan of Athens.

1910

- Electoral victory of the Liberal Party, under Eleftherios Venizelos, former minister of the Cretan State.
- Decree amending the plan of Athens, introduces the necessity of opinion by the municipal council.

- Law on taxation reform.
- Law on extraction of metal minerals and exploitation of mines.
- Suicide of a well-known author, admirer of traditional culture and critic of Athenian modernization. He rode on horseback into the sea.
- Plan for Athens by the German architect Ludwig Hoffman.
- Law on the Public Works Government Department, in charge of town planning.

1911

- New Constitution, with provisions for the protection of private property, except in cases of public benefit.

1912

- Law on municipalities and communes, abolishing the municipal act of 1833.

Balkan Wars of 1912–1913 and liberation of Northern Greece.

1913

- Assassination of King George I.

1914

- Foundation of the National Metsovion, Polytechnic of which the name is due to donors born in Metsovon, town of Epirus, and is still used today (called in English "National Technical University of Athens").
- Constitution of the Ministry of Transport, responsible for town planning.
- Law on the reconstruction of the town of Serres, destroyed by arson of Bulgarian forces during the 2nd Balkan War.
- Plan by the English architect Thomas Hayton Mawson for Athens.

1915

- Decree on the reforestation of Attica and Boeotia.

1916

- Conflict between liberals and royalists, national schism, and division into two states, i.e., the royalist state of Athens and the liberal state of Thessaloniki.

1917

- Devastation of the city of Thessaloniki by fire.
- One more plan for Athens, by Aristidis Balanos.
- Law on maximum building height.
- Law prohibiting building activity in the burnt area of Thessaloniki.
- Beginning of expropriation of large estates, mainly in Thessaly, for the rehabilitation of landless peasants.

1918

- Law on the implementation of a plan for Thessaloniki and of the instrument of "real estate groupings", a variant of land readjustment or reparcelling tools.

1919

- Royal decree on the land property documents required for the implementation of the Thessaloniki plan.
- Royal decree about the properties in the real estate groupings of Thessaloniki.
- Law on the resettlement of the population of Eastern Macedonia, displaced during the wars.
- Greek army units land in Izmir.

1920

- Law on gate trade tariffs.
- Treaty of Sevres, ratifying the Greek occupation of extensive lands.
- Unexpected defeat of the Liberals in national elections.
- Attempted assassination of the liberal leader Venizelos.
- Assassination of Ion Dragoumis, opponent of the liberals.
- Royal decrees on maximum building heights and on the implementation of the Thessaloniki plan.
- Law amending the previous decree on the Thessaloniki plan.

1921

- Law on the disposal of real estate properties in Thessaloniki.
- Decree fixing the area of the real estate grouping of Thessaloniki.
- Decree approving the street layout plan of section of Thessaloniki.

1922

Asia Minor disaster, Izmir is burning, and the national problem of refugees.

- Foundation of a health care fund for the refugees.
- Law for the promotion of manufacturing industry.

1923

The Lausanne Treaty reduces Greek territorial gains.

Population exchange, Commission for the Settlement of Refugees, and involvement of the League of Nations, which has the upper hand in the settlement process.

- Foundation of the Technical Chamber of Greece.

Legislative Decree of historic importance regulating the plans of towns, townships and settlements. Some of its provisions are still valid today.

2.3 Conclusion

The chronological tables which I presented in textboxes are an indicator of the chapters that will follow. They are a panoramic overview drawing attention to events, some of which will be described later in greater detail. These events are frequently of a purely social or political nature, but they provide the background of developments in the field of town planning. They may seem remote, but they had a definite impact on geographical space and on the growth of cities. Such effects also justify the reference to centuries long before Greek independence and to actions which ultimately shaped subsequent events.

Inter-relationships may be local, but also frequently international. National and local developments concern the political setup, social claims and antagonisms, even extraordinary situations, such as the spread of an infectious disease or a seismic event. But even such situations are intimately linked to the conditions of towns and settlements. The quality of housing and insanitary conditions are the link between an imported disease and the state of human settlements. At the same time political demands and military adventures impact on the internal financial environment and cause bankruptcies, social unrest and urban disintegration. Town planning is then reduced to a patching up operation and to an activity to satisfy the demands of a political clientele.

Continued conditions of dependence on foreign domination and assistance are soon translated into domestic political choices, which affect decisions transforming geographical space. Spatial planning, whether it concerns the town planning of a city, or the settlement of particular social groups, shapes the future of the territorial base on which social activity takes place. Hence, when the satisfaction of external demands takes precedence, particular governance practices are adopted, from law making to the regulation of urban land, through the medium of the transfer of foreign models or the influence of individuals, knowledge carriers, and alien ideologies.

The "planner" becomes the conduit of the importation of priorities, perceptions and techniques, whether he is the French "geometer" of the Capodistria period, the architect designing a plan of Athens under the Bavarian administration, the technocrat-engineer designing public works in the years of Tricoupis, or the anonymous bureaucrat of the League of Nations in charge of the settlement of refugees. The conceptual climate in which he is active is the product of ideas imported from countries in which he acquired his knowledge and experience, but also of compromises with the socio-political framework of the country where he now operates. To describe this complex set of interactions was my intention in this chapter and in the chronological tables. It will remain my goal in the rest of the book.

Chapter 3
The Remote Past and the Greek Independence War

Abstract In Chap. 3, the author tried to lend support to his claim that events of a very remote past, before the 1821 Greek War of Independence, or developments during the revolution itself, impacted on the work of post-revolution authorities. One has to return to the Byzantine and Ottoman periods to understand how custom-law and local power-sharing continued to influence town planning law and practice for several decades after independence. This explains references to the 14th century legal code of Armenopoulos or to the visions of national identity of Gemistos Plethon. Developments briefly listed in the chronological table of Chap. 2 are discussed in some detail. They include embryonic planning initiatives in Europe, Greek national assemblies during the independence war, the first constitutions, international protocols which determined the territory of the new state, and the controversies surrounding the distribution of national lands. Greek independence was secured in the 1827 naval battle of Navarino, when the fleets of the Protecting Powers (Great Britain, France and Russia) destroyed the Ottoman fleet. It then received the final seal of international legality in a series of protocols. In the meantime, the government of the country was entrusted to Count Capodistria, a Greek minister of the Russian Tzar (Count Ioannis Kapodistrias (Capo d'Istria) was born in the Greek island of Corfu. Capodistria is the version of his name habitually used in international literature).

Keywords Past centuries · Greek independence war · Protecting powers · National assemblies · Constitutions · International treaties

3.1 Introductory Note

After the end of the 1821 Independence War of Greece against the Ottomans, practically after the 1827 naval battle of Navarino, where the Allied Fleet of Britain, France and Russia destroyed the Turkish fleet, the independent Greek state entered a new phase. However, the previous centuries continue to weigh heavily on the country's social institutions, the Greek people's traditions, and the transitional administration's ability to cope with the needs of a ruined territory. There is a revolutionary

L. C. Wassenhoven, *The History of Town Planning in a Dependent State*,
https://doi.org/10.1007/978-3-031-96958-4_3

government of sorts, made up of coalitions of military chieftains and local notables, continuously riven by internal civil conflicts. Its ability to address the problems of towns and settlements is in fact nil. The country survives thanks to its traditional social structures and a judicial system with roots in Byzantine and custom-law. The management of towns is of course beyond the grasp of its revolutionary leaders, who ignore the vast amount of town planning practices followed in foreign countries which Greece would soon be called upon to imitate. But, when someone attempts to trace its history from a town planning perspective, the knowledge of this foreign experience, whatever its character, is an essential ingredient of his approach. Alien practices, whatever their applicability in a country exiting from centuries of foreign occupation and surviving at a marginal corner of Europe, were the inheritance of precisely those European countries which would soon play the role of "Protecting Powers" and would pull the strings of Greek government, in all its activities, town planning included.

3.2 The Tradition of European Town Planning

I shall return to the inheritance of the Byzantine era, with special emphasis on the first aspirations for a Hellenic Nationhood and on the persistence of custom-law, but I will first trace an outline of the town planning experience of major European countries which would soon find its way into Greek practice. The imported models were not of course limited to architectural designs of cities. They also included institutions like those of municipal government and a variety of legal texts and royal decrees. During past centuries there had been social events which affected history like seismic tremors, like the American and French Revolutions, which changed perceptions of the rights of citizens and of the concepts of private property, with inevitable repercussions on the nature of town planning.

The exact bearers of new ideas and practices are difficult to be identified, because, even though the Greek Independence War erupted only three decades after the French Revolution, these transmitters of ideas were not necessarily carriers of democratic principles. In independent Greece, when it practically becomes a Bavarian protectorate, with the blessing of the Protecting Powers, the messengers are German administrators, but also French military engineers. The active involvement of Greek architects and engineers educated in Germany and France is also an important factor, because of their exposure to well-known planning experiments in Europe, which may have been designed in third countries, like Britain, Spain or Russia. It is helpful, at this point, to attempt to provide a picture of these experiments, drawing on the existing literature and without of course pretending to write a full history of town planning.

Germany, long before the 19th century, had a tradition of a very active municipal administration, which included town planning affairs. Already in 1489, Munich had codified ordinances accumulated for centuries (Logan 1976, 379). On the eve of the Greek Independence War, in 1808, a Prussian law was issued, a sort of Town Charter, which regulated municipal administration. According to Ladd (1990, 15), the Prussian municipal *Selbstverwaltung* began with this act for the cities (*Städteordnung*), introduced by the reformist administration of Baron Karl von Stein, which became a model for later legislation. Town design continues of course to follow the model of princely 18th century traditions, e.g., in the early 19th century plans of Krefeld or Mülheim an der Ruhr, as well as that of Düsseldorf, work of the landscape architect Adolph von Vagedes. To be noted that the German states never allowed private construction initiatives to act freely. The Prussian government always insisted on preserving order and protecting public health (*op.cit.*, 89–90).

Although the empowerment and restructuring of local municipal authorities can be seen as a step toward the granting of town planning powers, equally important were the royal orders regulating traffic, construction, town limits and public health. According to Rossillion (1963, 207), a key component of town planning in 16th century France was the regulation of vehicle traffic. In 1539, king François I, convinced that the congestion of carriages and carts is causing traffic problems in the suburbs of Paris, issued an order prohibiting overtaking, excessive speed and direction reversals, except in street crossings. Davis (1966) adds that perhaps, during the Renaissance, Alberti or Da Vinci dreamed of ideal cities, but it was the law maker who laid the foundations of street building regulations, when streets were widened in accordance with the 1539 order, or when a 1534 decree on public health made street and pavement cleaning compulsory.

With the so-called de Sully decree of 1607 (Harouel 1993, 49) a number of town and street design issues were regulated: building projections on public ways, street width, and building architecture restrictions. Equally, this was the first time that economic and private property issues were taken into account. These problems appear again two centuries later in an 1807 law on the capture of surplus value, created by public works, or in an 1884 law on municipal plans of street alignments. Key motives were aesthetics and hygiene, as in an 1810 decree on unsanitary production activities (Rossillion 1963, 207; Davis 1966; Jacquot et al. 2019, 30), to which I will soon revert. Matters of straight-line street construction, house building and traffic policing had been dealt with in the 1607 decree of Henri IV (Gohier 1965). Because of the absence of town planning rules, the administration had to rely on street layout plans, which had to be respected by private initiatives to open new streets. These regulations remained in force even during the French Revolution, in accordance with a 1791 law, and also later under an 1807 law, as we shall see later.

Shortly after the de Sully decree, new orders were enacted in 1627 on the limits drawn between the city of Paris and the suburbs. In effect, this is an effort to tackle the problem of the expansion of the urban tissue, a task which is repeated in 1672. Kalflèche (2018, 10–11) speaks of the strenuous struggle to limit the size of large cities and the futile effort to prevent the expansion of Paris. In his view, the reasons of this effort were the fear of provisioning problems, the difficulties of maintaining law and order, and the care for public hygiene. As he explains further, a century later, in 1783, a decree had been issued, fixing a minimum street width of 10 m, requiring the stone paving of streets, and prohibiting buildings higher than 22.5 m. Wastewater had to be drained through underground canals and not middle-of-the-street open ditches. According to the wording of the decree, stone-paved pedestrian ways had to be constructed at such a level that would allow the pedestrians to walk along the external building walls, above the draining ditches. At about the same time, in 1794, a Prussian law was enacted, introducing a kind of early building regulation, to which I will refer in Chap. 5.

A new *Code Civil* is voted in France in 1804, under the shadow of constant ideological fluctuations. Land property is recognized as indivisible, i.e., not co-owned with the state (Schröteler-von Brandt 2018). The state is recognized as the apex of decision-making, in the sense of public power, but not as the supreme owner. With private property being substantially reinforced, urban development entered, in the 19th century, into a phase of limited planning activity and reduced public role. France was in a phase of transition from the pre-industrial to the industrial era (Bergeron and Roncayolo 1984). The modernization impulse, caused by the use of new technologies, the expansion of industry, and the cities' population growth, could not be checked by spatial policy means. Planning through private commercial relations subjected to land market forces was pursued increasingly by reaching compromises between public planning authorities and land proprietors. An imperial decree was published in 1810, with a long-lasting influence, secured a minimum of state control. As explained by Jacquot et al. (2019), it was the powers of policing real estate properties that gave the public administration the legal backing to intervene in urban areas and guarantee a minimum of sanitary conditions, security and aesthetic quality. The protection of hygiene was further enhanced by the power to police dangerous, inappropriate, noxious and insanitary activities. In addition, it offered to the administration the possibility to formulate a policy of industrial zoning, prevent the construction of newly appearing manufacturing plants in housing areas, and, finally, proceed to compulsory land acquisition for reasons of public interest (Gohier 1965). It was at about that time, in 1815, that the first effort was made to classify noxious activities, while earlier, as I explained above, the 1807 decree was published, concerning the servitudes caused by the recession of building lines and the capture of surplus values (*ibid.*). The French law on compulsory acquisition became a model copied by other European countries and a precedent of forms of participation by landowners, municipal interventions and compensation. It was these French laws that were implemented in French-occupied Rhineland, during the Napoleonic wars, and remained in force when this region was incorporated in Prussia in 1816 (Schröteler-von Brandt 2018).

In parenthesis, since reference was made to Germany, it is the right time to mention that it was in 1806 that the Kingdom of Bavaria was created, of which a prince was later to become the first king of Greece.

3.3 New Towns and Urban Interventions in Europe

The founding of new towns has been the object of a large number of studies by great historians of town planning. New town creation in the "modern age", to be carefully distinguished from the "contemporary period" (Wassenhoven 2019, 68–69), has been relatively limited, but even in the distant past there have been examples of great symbolic significance. One is tempted to recall how Ferrara, a splendid medieval city, had "new town" extensions twice in the 15th century (Benevolo 1980, 556–561). Palmanova, the prototype of the "ideal city", was built and fortified by Venetians at the end of the sixteenth century, for clearly defensive reasons (Lavedan 1959, 90–92). Its status as a "utopian" city in all its splendour, was, according to Burtenshaw, Bateman and Ashworth (1981, 13–14), the apex of "geometric authoritarianism" and influenced the French fortified towns of Vauban, to which I will return. In the modern age we find several examples of towns designed or renovated in the classical genre.

Following the views of Merlin (1994), we can single out a diversity of garrison towns, royal cities like Versailles (see also Phalip 2019, 36), port-cities, e.g., Rochefort, and early industrial centres, as the Royal Saltworks in Arc-et-Senans (Salines-de-Chaux), which included workers' housing estates and common-use spaces. We can include genuine new towns, such as Charleville and Richelieu in France, Karlsruhe and Mannheim in Germany, and Bath in England, notwithstanding some earlier urban development (Merlin 1994, 31; Lavedan 1959, 246–248). I shall have more to say about these urban developments. Noteworthy new town foundations in the 16th century, apart from Palmanova, were those of Coevorden in Holland, but also, according to Lavedan, Vitry-le-François and Villefranche-sur-Meuse in France, Valetta in Malta, and Zamosc in Poland (Lavedan, 1959, 76–92). We rediscover the model of the ideal city in Mannheim in the 17th century, during which several fortress towns appear in other parts of Europe (*op.cit.*, 248–266). Apart from the fortified cities, of great importance in France are Le Havre, Nancy and Charleville in the 16th century, and of the port cities of Rochefort, Brest and Lorient on the Atlantic and of Sète on the Mediterranean coast (*op.cit.*, 72, 93–100, 106–118 and 212–224). As to Karlsruhe, at the beginning of the 18th century, its plan follows a different tradition, but its geometric design, with radial streets converging to the palace square, inspired later imitations (*op.cit.*, 264–266). Let us not forget the case of Athens and the 1832 plan of Kleanthis and Schaubert, which followed a similar monumental design, worthy of the grandeur of the Bavarian king.

It would be a mistake to forget that the role of cities changed, already in the 18th but even more so in the nineteenth century, when, gradually, urban economic and social activities, care for hygiene and housing, and functionality, began to take precedence over defence, fortifications, and hegemonic symbolism. I shall return to these issues, after making some comments on the traditional prototypes which never ceased to be engraved in the subconscious memories of architect-planners.

I concentrate on only some of the initiatives which represent the centuries before the Greek uprising of 1821. One such case was the 1642 erection *ex nihilo* of the Ville de Richelieu, on a design of the architect Jacques Lemercier, which copied the model of famous medieval bastides (Kalflèche 2018; Joly 1991, 168; Steenbergen & Reh, 1996, 147). Of special interest are the city plans produced after a natural disaster. In London, after the disastrous fire of 1666, the famous architect Sir Christopher Wren submitted to King Charles II a plan for the Reconstruction of the City, which was not ultimately implemented (Lavedan 1959, 367–387). It should not escape our attention that Wren was an architect, astronomer, mathematician and physicist. A little later, in the 1680 s, we find cases of new towns for the settlement of refugees, as indeed it has happened frequently in human history in many countries, Greece included, in the 19th and 20th centuries. In Europe, the revocation in 1685 of the Edict of Nantes caused the settlement of Huguenots in Germany. A new suburban township was built in Erlangen for this purpose. In Kassel about 1,700 Huguenots were settled in the suburb of Oberneustadt. A town plan was produced in 1688 by the French architect, and Huguenot, Paul du Ry (Lavedan 1959, 467–468).

While the persecution of a religious minority was giving rise to these important planning experiments, in the same decade of the 1680 s, at the time of King Louis XIV and, initially, of his famous minister Colbert, France was encircled by the Iron Zone of fortified townships constructed by Marquis Vauban, a military engineer (Kalflèche 2018; Wassenhoven 2019, 146, 99). The territory within this zone became known as the French *pré carré* and the entire impressive, military and town planning project was the work of Marshall Vauban (Bitterling 2009, 138–153). In Vauban's perception, the entire space of France was a vast royal property, in need of management and protection, hence its enclosure in a *ceinture de fer*, a string-belt of castles. He convinced the French government that this zone, planned on the basis of detailed topographical maps, offered better protection than that of isolated fortified cities (Stoye 1988, 352–353). The new towns Vauban designed and built, mostly on river shores – Saarlouis, Fort-Louis, Neuf-Brisach and Mont Royal on the Rhine—remain famous (Lavedan 1959, 224–228).

At the turn of the century, from the 17th to the 18th, a major town planning event was the creation of a new European capital, Saint Petersburg. For a long time, it remained the seat of the court of the Russian Tzars, hence it was there that Count Capo d' Istria (Capodistria), a nobleman of Greek extraction, offered his services to the Russian Emperor, before being appointed as Governor of independent Greece in 1828. In the belief that his memories must have influenced his views on city planning, I shall return to the subject in the next chapter. Moving now to a totally different kind of town, I can mention the planning interventions taking place during the same period in Bath, the famous English spa. The planning and architectural design of European spas is of interest for the Greek case, since relevant plans are produced during the reign of the Bavarian King Otto, and the first spas are inaugurated later in the 19th century. For Bath a plan was produced in 1724 by the architect John Wood (Lavedan 1959, 457–462), who also designed several buildings, in a period when Bath was a most fashionable, high-society destination.

Having mentioned Saint Petersburg, and having in mind the planning of Athens 65 years later, I turn to the extension of another capital, viz. to the building of New Edinburgh, designed in 1767 by the Scottish architect James Craig (Lavedan 1959, 462-0.467). Peter Hall refers to a New Town, the flower of town planning before the industrial revolution (Hall 1975, 20–21). To be noted that in 1765 Craig had won the first prize in a competition for the plan of the new town.

With regard to industrial settlements, I repeat the example of Salines-de-Chaux, which I mentioned earlier, near Besançon, designed in 1774 by Claude-Nicolas Ledoux (Ragon 1986, 74). It is an early example of the Age of Enlightenment, which follows a rational geometry and a hierarchical structure of the project's components.

The plan was designed initially without reference to a specific site and presented to King Louis XV. In the eighteenth century, plans are also produced for existing French cities, e.g.., Marseilles, Arras, Rennes and Limoges, often accompanied by regulations on aesthetics and traffic limitation (Kalflèche 2018).

Moving to the other side of the Atlantic, we find the work of another French architect, Pierre Charles L'Enfant, for the 1791 plan of Washington, new Federal Capital of the United States of America (Wassenhoven 2019, 58–59). L'Enfant had previously taken part in the American War of Independence against the English. Washington D.C. was not of course a town founded on foreign soil by a colonial power, like New Delhi in India, but its roots were two English colonies, Georgetown (Maryland) and Alexandria (Virginia). After independence, the 1790 Residence Act referred to a federal capital on the River Potomac, on a site selected by George Washington himself. Before the Georgetown trading station, the Dutch had founded their own station in present-day Manhattan, under the name of New Amsterdam (Flitouris, 2015, 107). After 1660, New Amsterdam was captured by the English, and renamed New York, which after the American independence served briefly as capital of the United States (Brunn & Williams, 1983, 66). Let's mention, as an amusing interval, that in the provisional government gazette of the Greek revolutionaries in the 1820s, New York is mentioned in Greek as … *Nevorakon*! Further north, in Canada, the city of Quebec was founded by the French in 1608 (Flitouris 2015, 98). Going back to the days of the planning of Washington, in faraway Russia of the Empress Catherine the Great, a new plan was drawn in 1794 by the Flemish engineer Franz de Voland for the port of Odessa, known as the pearl of the Black Sea (Wassenhoven 2019, 74–75· Brunn and Williams 1983, 128–129).

The age which I explore has not bequeathed us only town plans, of existing or new cities. Of importance are also urban renewal interventions for the opening of free spaces, plazas and boulevards, which were imitated later. I will mention a few, but famous, examples, beginning with the royal Plaza Mayor in Madrid, of which the construction took place in the 1590 s, but was continued for several years in the reign of Philip III, by different architects (Lavedan 1959, 413–416). Today, its final form is attributed to the architect Juan de Villanueva, who redesigned it after a fire in 1790. Two more illustrious examples are two Paris squares. The first is the Place des Vosges (former Place Royale), designed in 1605, in the days of Henri IV, who intended to settle Italian silk weavers (Rossillion 1963, 228). The second, but equally famous, is Place Vendôme (Benevolo, 1967, 13; Lavedan 1959, 503). Although opened much later, we can include in this category, the cases of Regent's Park and Regent's Street in London, opened in 1810 on the basis of plans by John Nash and of J. and D. Burton (Abercrombie 1967, 75). There are. no doubt. many more examples of open spaces and avenues, which most certainly influenced the imagination of architects who were invited to design cities in third countries, including those of Greece after its independence.

3.4 Byzantine Era, Custom-Law, Ottoman Empire and the Greek Diaspora

For centuries, there was no way for this European experience to be transmitted to a country ruled by the Ottoman Empire. Such territories were governed by local beys, in accordance with *firmans* (decrees) issued by the Sultan. However, at the local level (Inalcik 1995) justice administration was often handled by elders, who followed custom law, inherited from the Byzantine Empire. This was especially the case when local disputes concerned land property and inheritance. In independent Greece, albeit territorially still handicapped, the effort to implement a sort of land policy had to rely on legal codes which were yet unavailable. Custom law had to be respected, which in traditional communities was based on the 14th century codification of Byzantine law by Konstantinos Armenopoulos, known as *Hexabiblos* (Veremis et al. 2018, 27). Armenopoulos was a jurist and served as a judge in Thessaloniki and his code was divided into six books with civil and penal law sections, as well as an annex of rural law. Armenopoulos had drawn material from the Emperor Justinian's legislation, as well as from Roman law. His code was still applied in Balkan countries in the 19th century (Vasiliev 1954, 886–887). In his study, Nakos (1991, 187–190) reaches the conclusion that the direct or indirect application of the Hexabiblos in Greece continued throughout the period 1835–1946. As far as the period of the Bavarian government is Greece is concerned, this is confirmed in the writings of both Georg Ludwig Von Maurer, member of the Bavarian Regency, and his assistant Karl Gustav Geib, both professors of law in Germany (Maurer, 2021, vol. II, 254–255, 303; Geib, n.d., 97–98; Sakellaropoulos 1991, 153). In fact, Geib detects in the Hexabiblos the first signs of the concept of pre-emption rights, although it rather takes the form of the right of the nearest relative or neighbour, applied in the cases of sale of landed properties in the absence of rightful inheritors (Geib, n.d., 97–98). According to Geib, property mortgaging and the noxious impact of adjacent onerous activities, e.g., threshing, also followed custom-law. Mortgaging was often described by the term *amānāt* in Arabic (Gerontas 1969, 502).

Byzantine inheritance left a mark of even greater importance. In the 14th century years of the decline of the Byzantine Empire, semi-independent provinces emerged of which the Despotate of Moreas (or of Mystras) is of particular interest. The court of the despot in Mystras was compared to those of Italian Renaissance Princes (Vasiliev 1954, 793). It was there that the seeds of a purely Greek nation were sown, mainly by the philosopher and humanist Gemistos Plethon. He was convinced that Hellenism can be reborn in Moreas, the present-day Peloponnese. He also wrote about the social innovations that should accompany his vision. Plethon designed a radical reconstruction of the social system and of the agricultural economy. In his view three social classes could be distinguished, i.e., rural peasantry for land cultivation, craftsmen for the manufacture of implements, and the class of administrators and soldiers. He envisaged an army made only by pure Greeks. He also differentiated the tax payers from the soldiers, who were exempted from taxation. All the others were helots. Plethon rejected private land property. In his opinion, all land is common property and every individual has the right to build his house and cultivate the land anywhere he chooses. Some commentators find similarities between the views of Plethon and those expressed in Rousseau's *Contrat Social*, or the ideas of Saint-Simon (*idem.*, 794–796). It is clear that Plethon was dreaming of a Hellenic state in the Peloponnese, where a specific land policy would be applied, including extensive land readjustment and nationalization. A sign of the unity of this state was the decision of the Byzantine emperor Manuel, father of the then despot of Moreas, to build a defensive wall along the Corinth Isthmus. Soon, the Byzantine Empire reached its end after the city of Constantinople was captured by the Ottoman forces in 1453.

During the long centuries of the Ottoman period, present-day Greece went through different stages which were influenced by the gradual decline of the Ottoman Empire. It was only a few townships, which developed trading relationships with the outside world, that prospered in the 18th century (Moskof 1972, 66–84). Athens was a special case of a walled centre, surrounded by an external space of plateaus, *mahallahs* (neighbourhoods), and parishes (Karidis 1981, 226–243). A complex of *mahallahs* constituted a plateau, which roughly corresponded to an administrative unit. Karidis' estimate is that in mid-17th century Athens had about 10.000 residents, distributed among 8 plateaus and 32 neighbourhoods. The protected centre was the residence of the ruling class (see also Karidis, 2014). Athens was the scene of fierce battles during the independence war and at the time of liberation was in fact a mere large village (Fig. 3.1).

Although Hellenic nationalism and the existence of various forms of feudalism have a pre-Ottoman past, the situation under the Turks gradually takes a new shape from the view-point of the peasant share-cropper. A version of feudalism prevails,

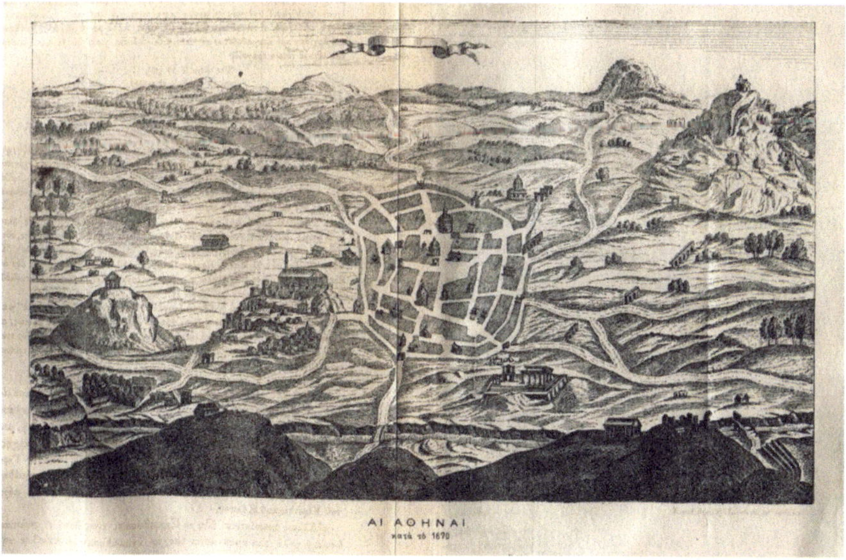

Fig. 3.1 Athens: engraving of 1670. *Source* Kampouroglou (1889, 48)

although different from that of the West-European medieval variety. Slowly, it is the Greek notables and *kodjabashis* that gain the upper hand in the allegedly self-governed communities. As the decadence of the Ottoman Empire continues, the imperialist penetration of the West undermines Ottoman power, at a time when trading agreements, the famous capitulations, signed by western countries and the Sultan, facilitate the rise of a class of Greek merchants and shipowners, working under a foreign flag and enjoying special privileges (Wassenhoven 2019, 158). This is a development of great importance for the growth of the urban network of Helladic space, especially of port-cities in the islands. Although the first capitulation agreement between the Ottomans and the French was signed as early as 1535,[1] of much greater impact was the 1740 treaty. As explained by Olson (2010), the treaty included 85 articles, most of which concerned commercial, economic and legal capitulation. This was the peak of French influence in the Ottoman Empire. In the following years, the French had no serious competitors in trade and transportation, even between Turkish ports. Later of course similar treaties were signed between the Ottomans and England, Holland and Prussia, in an effort to balance French influence.

Capitulation treaties favoured Greek entrepreneurs, who were able to operate under the protection of European powers, to which they owed allegiance. As explained by Beaton (2020, 159), the Greeks were able to occupy positions of power in trade and the banking system. The main banks of Istanbul were owned by Greeks. Even the Sultan's Court had banking accounts in Greek banks, which were the frontline of modernization. The enormous advantages earned through the capitulation agreements were beneficial for the class of merchants and shipowners not only of the Aegean islands, but also for the entire Greek Diaspora. The wealth and political power enjoyed by the members of Greek communities of various European and Middle East cities were of great importance for the rise of a nationalist ideology, particularly if we take into account the absence of a genuine urban middle class in the territories of later free Greece. The international Greek Diaspora was the front guard of the *heterochthons*, i.e., the foreign-born Greeks, in the new state, who soon became the opponents of locally-born *autochthons*, in other words, of local primates (Clogg, 2021, 47; Vakalopoulos 2005, 19). The Greeks living in London, Odessa or Trieste, had a totally different understanding of the concepts of statehood and nationhood from that of the autochthons. Before the revolution, Gallant remarks, there was no nation, let alone a people with uniform national conscience (Gallant 2017a, 84). There was of course the sense of belonging to a Turkish religious *millet*, of speaking the Greek language, and of being a Christian Orthodox (Clogg 2021, 10). The fundamental feeling was rather the sense of belonging to a local community. It is to the Greeks of international communities that addresses his nationalist call the author "Anonymous the Greek", in his historic treatise "Hellenic Nomarchy, Speech on Liberty" of 1806, where he invites them to develop their activity on Greek soil for the benefit of the nation (Anonymous the Greek, 1980, 152–153).

Of even greater importance than the French-Ottoman capitulations, was the 1774 Russo-Turkish Treaty of Kuchuk-Kainarji, following a Russo-Turkish war, which

[1] See the Wikipedia lemma "Capitulations of the Ottoman Empire".

guaranteed Russian protection of the Christian citizens of the Ottoman Empire (Moskof 1972, 84–87; Veremis et al. 2018, 30; Vournas 1974, 92). Greek shipowners and sea-captains could now hoist the Russian flag and build ships of increased tonnage.[2] Russia was of course caring for its geostrategic interests, but for Greeks this was an incentive to settle in Odessa, where Empress Catherine the Great provided land for Greek colonists. Another Russo-Turkish war ended with the Bucharest Treaty of 1812 (Vournas 1974, 92).

As nationalist aspirations are on the ascent, apart from the invitation by the anonymous writer urging all Greeks to return to their homeland, Count Capodistria, senior official in the Russian government, described the Greek situation in an 1811 memorandum. The future Governor of Greece distinguished four social classes: The land owners and the aristocracy, either of the Fanar district of Istanbul, or of the Hospodars' clan of Wallachia and Moldavia; the Clergy and the intellectuals; the merchants and shipowners; and the workers and warriors (Ploumidis, 2014, 79). A major event was coming closer, with the founding of the Friendly Society in Odessa by members of the Greek Diaspora, which was organizing an insurrection in the Balkans. It failed because of the conflicts between opposing factions, possibly also because of their inability to mobilize the lower classes (Dakin 1972, 44).

3.5 The years of the Greek Revolution

The Greek Revolution finally erupts in 1821, at a time when several insurgencies take place in other countries too, either against oppressive regimes, or against foreign occupation, as in Italy of which large territories were under the Austrian yoke. It is curious, but quite natural, that the first country to salute the Greek uprising was the faraway Haiti, which had recently acquired its freedom (Vournas 1974, 94). However, conditions were totally different from those of other European countries, where the first signs of the industrial revolution had appeared or, quite simply, prevailed a system of economic organization, based on international trade and advanced urbanization. On the contrary, in the geographical space which was to reach independence in 1832 with the blessing of the Protecting Powers, urban development is still absent. Fragments of urbanity can be discerned in settlements of shipping and commercial communities, but these lack the traits of cities in advanced European countries. Education levels are low and the dissemination of knowledge and information is problematic. In spite of the decline of the Ottoman Empire, described elegantly as the Eastern Question, the centres of Helladic territories are underdeveloped and a social and demographic crisis is evident (Beaton 2020, 49, 98; Vakalopoulos 2005, 259). The revolution itself accelerated the crisis. According to Beaton (*op.cit.*, 75), between 1821 and 1828 the regions in revolt lost about 29% of their population (Fig. 3.2).

---

[2] See the Wikipedia lemma "Treaty of Kuchuk-Kainarji".

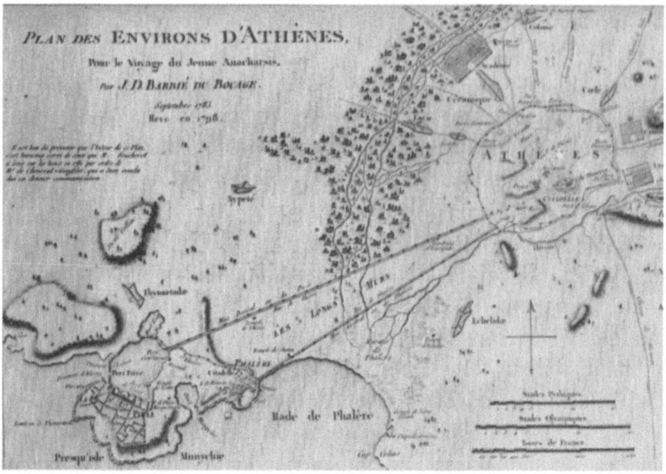

Fig. 3.2 Athens and Piraeus in 1798. *Source* Kampouroglou (1969, vol. III, p. 70)

At the beginning of the revolution, very few townships can be remotely described as urban (Kotea 1999; Loukatos 1978). The case of Athens is slightly different. The structure of these townships hardly changed in the course of the revolution. As Kostis (2018, 13) put it, the state becomes a reality in 1830, but its "urban" centres still reflect a pre-revolution model. In these centres the dominant force is not of course the bourgeois class in the western sense, but rather the primates, notables, *kodjabashis* and aldermen, who hold power as the Turkish agas did earlier. It is these primates who convene the first assembly and constitute the Messenian Senate, and later the Peloponnesian one. They draft a manifesto and seek the assistance of the Christian states. It has been described as the first legal document of the revolutionaries (Vournas 1974, 79–80), which inspired the later first Constitution of Epidaurus (Beaton 2020, 94), The text reached the philosopher Jeremy Bentham who sent to the provisional government a letter full of praise. Soon arrive in Greece two key figures, Prince Demetrius Ypsilantis, a Phanariot,[3] military officer in the Russian army, and son of the Hospodar of Moldavia and Wallachia, and Alexandros Mavrokordatos, also a Phanariot, who spoke 8 languages, had acquired a great deal of political experience in the court of his uncle, a Wallachian prince, but had never visited Greece before the revolution. Both played a great role in the course of the revolution and the first years of the new state. Ypsilantis was perceived as a leader who enjoyed Russian support, which made him very popular. Mavrokordatos was a subtle politician, who soon realized he had no supporters in Moreas, and made Central Greece his political base. In 1825 he was already Secretary General of the government's Executive (Dakin, 1972, 45).[4] Some Marxist writers consider him the "evil demon"

[3] From Phanar, quarter of Constantinople.

[4] A great deal of information used in this or the subsequent chapter was collected from the General Gazette of Greece, the contents of which can be found in the data bases of the General Archives of

of the revolution and a supporter of British interests, but others, of equally Marxist inclination, describe him as the revolution's most important figure. Veremis et al. (2018, 61–62) consider him as a most-important link with philhellenic circles, an extremely competent negotiator, a skilful manager of international loans, the main author of the first Greek Constitution, and a victim of populist attacks. Petropoulos (1997, 123–124) adds that Mavrokordatos believed that personal preferences had no place in state affairs. He was not dogmatic in his views, was rather a realist, hence he did not hesitate to exploit the antagonism of Britain and France, and knew that ultimately the British priorities would prevail.

The activity of Mavrokordatos is a good example of the division between the heterochthons, like him, and the autochthons, well represented by Petrobey Mavromichalis, leader of a powerful family clan in Mani. This division was about political power, but it also had serious effects on local affairs in towns, of which the land becomes a terrain of dispute, since its possession was now the object of desire of both locals and outsiders. Town planners were soon called upon to reconcile such claims and disputes. Vakalopoulos (2005, 198–199) recognizes the justified complaints of local citizens who had fought in the war and faced serious problems of poverty and survival. On the other hand, he comments, the heterochthons, regardless of their country of origin where they had been educated and/or made a fortune, were now occupying key posts in the government or were engaged in speculative activities in independent Greece. Conciliatory solutions were found later with regard to the possibility of the foreign-born to be appointed in the public administration.

Divisions did not only exist between autochthons and heterochthons. Dividing lines separated even the power elite. According to Petropulos, there existed three ways of classification in distinct categories: (a) military leaders and local politicians, i.e., primates; (b) according to region of origin; (c) ideological convictions, i.e., modernizers, western values' sympathizers, and carriers of traditional, autocratic values (Petropulos 1968, and Petropoulos 1997). Veremis, Koliopoulos and Michailidis (2018, 58–60) remark that landless peasants who had been recruited by local chieftains to fight in the war, often against their will, had only one demand, to be given a land parcel for cultivation. Naturally, they viewed the arrival of heterochtons with suspicion.

The resulting socio-political landscape was fluid and unstable. Alliances and divisions were constantly fluctuating (Clogg 2021, 35–39). Ideological and practical conflicts repeatedly appeared and disappeared depending on the circumstances. The poor peasants were not only diffident and suspicious of the real intentions of the new elite, but also of the local primates. An old fighter and adjutant of General Kolokotronis had apparently said that the new-style *kodjabashi* differed from the old Turkish aga in one respect only: He was called Yiannis instead of Hassan and attended an Orthodox Church instead of a Muslim Mosque. However, local attachment remained the dominant identity of the local population (Dertilis 2019, 149–150). Local language idioms were a reflection of regional origin. Communication

Greece, the Greek Parliament and the National Academy archives. Precise references are not given to avoid excessive detail.

was frequently a problem. Dertilis rightly recalls the theatrical play *Babylonia*, by D. Vyzantios, first performed in 1836.

In the middle of this fluent environment, foreign philosophical currents reach Greece. 1821 is not only the year of the revolution but also of the publication of the book by Saint-Simon on the industrial system. Before its translation in Greece, foreign adherents of Saint-Simonism arrive and occupy posts in government, after independence, like the Frenchman Gustave Séligmann d'Eichtal, who is appointed as advisor of in the Directorate of Economics. Other Frenchmen are also appointed in various departments, all eager to promote a new approach to development, to encourage infrastructure works, and the creation of model agricultural farms (Moskof 1972, 174–175). Saint-Simon, who grew up in the turbulent years preceding the French Revolution, was preaching the increased role of the artisans and workers. His views are extensively presented in textbooks of town planning history precisely because of this ideological position on production, efficiency and labour (Benevolo 1967, 54–55; Ragon 1986, 58). Saint-Simonism penetrated the mind of perhaps only few individuals, but certainly of those with an inclination to adopt new positivistic models of order and organization (Moskof 1972, 177), which were bound to affect town planning.

3.6 National Assemblies, Foreign Influence, and International Treaties

In this environment of a power struggle of interests, an ideological fermentation takes place, manifested in a series of National Assemblies, Constitutions and decrees of dubious impact. It is in this context that a debate develops on urban policy, new town creation and town design. However, a policy of whatever subject relies on the existence of legal and organization structures, which were at an infancy while the war was still going on. A new Constitution was supposed to be the foundation of the state. I am not going to write a constitutional history of Greece, as several professors of law have already done (Alivizatos 2011; Marias and Pantelis 1996; Kaltsonis 2017). The first Constitution was approved by the National Assembly of Epidaurus, convened in a village near ancient Epidaurus in 1822, the year of the massacre in the island of Chios, which inspired the poem "The Child" by Victor Hugo.

Les turcs ont passé là. Tout est ruine et deuil.
Chio, l'île des vins, n'est plus qu'un sombre écueil.

A question is addressed to the Greek child:

Que veux-tu ? fleur, beau fruit, ou l'oiseau merveilleux ?
– Ami, dit l'enfant grec, dit l'enfant aux yeux bleus,
Je veux de la poudre et des balles.

It also inspired the painting of Eugène Delacroix "The Massacre at Chios".

Beaton (2020, 87–88) refers to the successful effort of Mavrokordatos and his followers to include in the Constitution the principles of a centralized state, law enforcement, and democratic responsibility. For Vournas (1974, 107), the Constitution did nothing but confirm the power of the oligarchy. In the opinion of Sakellaropoulos (1991, 38–39), the Epidaurus Constitution and that of Troezen in 1827 put an end to serfdom, imposed the liberty of movement, and sanctioned the freedom of contracts. In comparison to other European countries which were struggling for freedom, Greece was at the front line.

Local administrations and senates were being created during the Revolution in various regions of the country. At the national scale, the Epidaurus assembly created a parliamentary body and an executive authority, under a chief secretary and 7 ministers, but failed to reach a decision on the issue of national land distribution, at a time when the local primates were busy buying land in auctions. They also jointly dispatched emissaries to West European countries to find a king for Greece and negotiate an international loan.

An 1823 National Assembly at Astros allowed the sale of national land to finance the war effort, with the result that local primates managed to buy legally land abandoned by Turks who were leaving Greece (Vournas 1974, 125; Kostis 2018, 311). Internal fighting between opposing factions continued, soon to degenerate in 1824 into open civil war (Alivizatos 2011, 48–49; Vournas 1974, 128–140; Beaton 2020, 90). It sadly coincided with the massacres of the islands of Kassos and Psara. For a while Greece had two governments and two parliaments.

Shortly after, in 1825, the Greek provisional government submitted to Great Britain a petition to become the sole protector of Greece. It has been described as the Act of Submission (or Subjugation). Its rationale was explained in an open letter by the distinguished politician Spyridon Tricoupis. Simultaneously, the Francophiles addressed a request to the Duke of Orleans, seeking his consent for the crowning of his son Duke of Nemours as King of Greece (Dakin 1972, 53; Kaltsonis 2017, 37). The French General Roche had arrived in Greece to negotiate this option, to bypass the English choice of Prince Leopold of Saxe-Coburg (Petropoulos 1997, 120–121). Notwithstanding these negotiations, the English, French and Russian overlords agreed to act as mediators between Greece and the Ottoman Sultan (Beaton 2020, 99–100). Part of the French infiltration was Roche's offer to the Greek authorities of a collection of French legal documents, laws and decrees of the period 1789–1814, which would serve as models. A great deal of international mobility can be observed in those days regarding the future of Greece. A Bavarian senior officer, later member of the Regency of King Otto, arrives in Greece, and distinguished Bavarians sign an appeal for support to Greece. Among the signatories was the Bavarian Court architect Leo von Klenze, later town planner of Athens. The diplomat of the Russian Court Count Capodistria, born in Corfu, visits Paris, London and Brussels for negotiations. In 1827 he was selected by the Troezen National Assembly as Governor of Greece. This assembly appointed an interim government of which the general secretary was G. Glarakis, a physician born in Chios and educated in Germany and Austria, who was later to occupy several ministerial posts.

The "hard" events, as always, were those taking place in the battlefield. A joint English, French and Russian fleet reached Greek waters, commanded by the Admirals Codrington, De Rigny, and the Dutch Van Heiden, and soon destroyed the Ottoman-Egyptian fleet at the naval battle of Navarino. Three streets in central Athens bear their names. The Egyptian force under Ibrahim Pasha, which in the meantime had turned the Peloponnese to ruins, hastened to return to Egypt. The Protecting Powers soon signed the first London Protocol which sealed Greek independence (Dakin 1972, 54–58; Vournas 1974, 201). Repeated antagonistic national assemblies (pro-Russian or pro-British) were in the meantime continuing. The Anglophiles managed to appoint two English senior military officers, Richard Church and Alexander Cochrane. The Russophiles were pleased with the appointment of Capodistria as Governor. The Francophiles were the losers (Petropoulos 1997, 128–129). Worth mentioning is that the 19th century historian Mendelssohn-Bartholdy considered the choice of Capodistria as a coup-d'état (Mendelssohn-Bartholdy, 1876, 27). Yet the Troezen Constitution, in spite of compromises, is considered as genuinely democratic and inspired by the American Constitution of 1787 (Kaltsonis 2017, 39–41). During the same period, two decrees reiterated the will to distribute land to fighters, widows and orphans (Kostis 2018, 312), but the government was not in a position to control domestic conditions in a country plagued by hunger, larceny and piracy (Beaton 2020, 93). The real achievement of the country, due to perspicacious politicians, educated Greeks, merchants, and shipowners, was that independence acquired an international dimension, even to the detriment of national unity.

3.7 National Lands, Towns and Town Planning

The reference to the Troezen Assembly, helps me to switch to the problems of towns and town planning, because it voted a decree allowing Greeks from Izmir to found a new town, by the name of New Smyrna, near the Corinth Isthmus. Initiatives for the creation of new towns often had a colonialist intention. I refer to the proposal of 1823 by the English Colonel Stanhope for the colonization of the country by Englishmen or other foreigners, able to promote agriculture or staff the army and the administration (Themeli-Katifori 1985, 99). Involvement of foreigners in town planning was due to become frequent in the following years. A French officer, probably Colonel Charles Nicolas Fabvier, designed the central avenue of the town of Kalamata in 1825 (Lavedan 1952, 199). The role of Greek and foreign engineers and army officers became very prominent later, but their presence was noticeable even in the course of the revolution. Theodoros Valianos, captain in the engineering corps of the Russian army, is credited with repair and rearrangement of a building in Nafplion, which became the seat of parliament. Fabvier deserves a further comment because he was a veteran of the Napoleonic army, became commander of a Greek army unit, and was one of those in favour of creating an industrial and agricultural colony of foreigners. He was an adventurer, like the British Richard Church, Thomas Lord Cochrane, and Frank Abney Hastings, who was fired from the Royal

Navy for reasons of disobedience, but was then distinguished as commander of a Greek battleship. Such officers were not mercenaries, but were escaping from the security services of their own countries (Beaton 2020, 95). Their case was different from that of anarchists, e.g., Italian Carbonari, who joined the Greeks for ideological reasons (Vournas 1974, 95–97, 149).

Individuals like Fabvier or the British officers had offered their services to the primitive Greek fighting units. But this is a period when Greece was badly in need of assistance for the care of the wounded soldiers or of patients suffering from diseases like cholera or because of malnutrition. The creation of hospitals was a priority and the assistance of foreign physicians, like the French surgeon Bailly and the German Treiber, was greatly appreciated. Health care was a field requiring a consistent public policy, and so was law enforcement because of widespread banditry, piracy and illegal trade of antiquities. The 1825 National Assembly voted for the creation of a 4-tier system of law courts, of which the structure is basically maintained even today. The issue that causes most controversy is that of national lands, that now belong to the emerging state by virtue of the right of war. In 1825, the nascent parliament authorized the sale of part of these lands and appointed a committee to manage the procedure. Ironically, a member of the committee was later the purchaser of huge tracts of land in Attica. A little later, in 1826, a new assembly reversed the decision to sell national lands and the issue remained open. The same assembly ratified the decision to declare Britain as the protector of Greek interests. All these developments were, however, overshadowed by the Exodus of the citizens of the besieged town of Messolonghi, where Lord Byron, an ardent Philhellene, had died a couple of years earlier.

Successive assemblies were convened in various places in the country, often by dissenting political factions, including that of the island of Aegina, which later became the first capital under Count Capodistria. Athens is still far from having a dominant position in the national network of cities, but its siege by Turkish forces in 1826 and the bombardment of the ancient Acropolis make it the centre of attention. The provisional government announced that the fortress of Athens is at its deathbed. General Karaiskakis, leader of the Greek forces, was killed in battle and Athens was occupied by the Turks in June 1827. While existing towns experience a great deal of destruction, the interest for the building of new settlements to accommodate refugees, Greeks of the diaspora, or other settlers never ceases (Kafkoula et al. 1990, 89), in a country which is devastated and lacks basic infrastructures and services. The development of an elementary road and railway network would have to wait for several decades, when in England, the first railway lines become operational in the second half of the 1820 decade and in France the first line is inaugurated in 1831 (Ragon 1986, 25). Town planning interventions multiply on a European scale, including urban renewal projects in Munich designed by the architect of the Bavarian royal court Leo von Klenze, who later produced a plan for Athens. The design of the first town plans in Greece would have to wait until the government is in the hands of Count Capodistria, as I will discuss in the next chapter.

3.8 Conclusion

This chapter does not deal with the period of independent Greece and, consequently, does not refer to town planning laws, town plans, and urban development efforts. Yet the 1820s are full of concepts and ideas which appear on an international scale and are hesitantly transferred to Greece. The main land-related problem in Greece was of course the future of national, state-owned lands now occupied by the revolutionary forces. The issue of distributing land to specific groups and the creation of settlements for refugees and newcomers was hotly debated. These were problems due to remain open in the following years.

The period of the insurrection left behind wounds which had to be healed. Human losses in battle, or because of hunger and disease, were enormous. Material damages and shortages of shelter required urgent attention. Infrastructures and public facilities did not exist. These were the priorities of an elementary rehabilitation programme which subsequent administrations had to devise. Yet, the problem was not simply of a material nature. It also embraced power structures, either inherited from the past, or created by foreign pressure. The struggle of local power holders and incoming innovating individuals of the Greek Diaspora was to continue. New arrivals were not necessarily motivated by patriotic sentiments, but also by profit-seeking pursuits. These power struggles may seem of secondary importance when compared with the plight of poverty and destruction, but they are the background of political developments and priorities, with immediate effect on town planning activity.

An important aspect of power antagonisms is the distribution of decision-making responsibilities at national, regional and local level and the staffing of administrations with the right personnel. The character of this problem was evident already in the years of the independence war, when the provisional government had to rely on foreign military engineers and medical personnel. The availability of such experts was not only a matter of assistance by philhellenes, but also of the presence of foreign military missions of the Protecting Powers. The expertise transplanted to Greece was accumulated in their country of provenance and was then incorporated in solutions of various varieties, including those of a town planning nature. However, these solutions may or may not be compatible with pre-existing social practices, which were regulated by traditional wisdom and custom-law. These problems did not disappear overnight. They were due to preoccupy the administration of Count Capodistria and that of the Bavarian government.

References[5]

Abercrombie P (1967) Town and country planning. In: Rigby Childs D. Oxford University Press, London (originally published 1933)
Alivizatos N (2011) The Constitution and its enemies in modern Greek history, 1800–2010. Polis, Athens [GR]
Anonymous the Greek (1980). Hellenic Nomarchy, i.e. Speech on Liberty, 1st edn., 1806. Kalvos, Italy, Athens [GR]
Armenopoulos K (1971) An outline of laws (*Procheiron Nomon*) or Hexabiblos. Edited by KG Pitsakis. Dodoni, Athens [GR]
Beaton R (2020) Greece: Biography of a modern nation. Penguin Random House, London
Benevolo L (1967) The origins of modern town planning. Routledge & Kegan Paul, London (Italian original 1963)
Benevolo L (1980) The history of the city. Scolar Press, London (Italian original 1975)
Bergeron L, Roncayolo M (1984) De la ville préindustrielle à la ville industrielle, essai sur l'historiographie française. Greek translation. EMNE-Mnemon, Athens (French original 1974) [GR]
Bitterling D (2009) L'invention du pré carré. Albin Michel, Paris
Brunn SD, Williams JF (1983) Cities of the world. Harper & Row, New York
Burtenshaw D, Bateman M, Ashworth GJ (1981) The city in West Europe. Wiley, Chichester
Clogg R (2021) A concise history of Greece. Cambridge University Press, Cambridge
Dakin D (1972) The unification of Greece 1770–1923. Ernest Benn, London
Davis J (1966) Zones à Urbaniser en Priorité et Zones d' Aménagement Différé. Conférence. Ministère de l'Équipement, Paris
Dertilis G (2019) History of modern and contemporary Greece, 1750–2015. Revised edition. Crete University Press, Herakleion [GR]
Flitouris LA (2015) Colonial empires. Asini, Athens [GR]
Gallant TW (2017) Modern Greece: From independence to the present, 1st edn., 2001. Bloomsbury Academic, London
Gallant TW (2017). Modern Greece: From the independence war to our days. Greek translation. Pedio, Athens (English original: Modern Greece: From the war of independence to the present) [GR]
Geib G (n.d.) An account of the law conditions in Greece during the Turkish rule up to the arrival of King Otto I. Greek translation (German original 1835). In: Avdi-Kalkani I (ed) (n.d.). The account by Dr Gustav Geib of the law conditions in Greece. Govostis, Athens, pp 57–170 [GR]
Gerontas D (1969) History of the Athenians, vol. IV. Palmos—N. Antonopoulos, Athens [GR]
Gohier J (1965). L'évolution de l'urbanisme en France. CRU65 - Études et Essais. Centre de Recherches d'Urbanisme, Paris
Hall P (1975) Urban and regional planning. Penguin, Harmondsworth
Harouel J-L (1993) Histoire de l'urbanisme. Presses Universitaires de France—Collection Que Sais-je?, Paris
Inalcik H (1995) The Ottoman Empire: The classical age—1300–1600. Greek translation. Alexandreia, Athens (English edition 1973) [GR]
Jacquot H, Priet F, Soazic M (2019) Droit de l'urbanisme. Dalloz, Paris
Joly R (1991) La ville et la civilisation urbaine. Greek translation. Synchroni Skepsi, Athens (French original 1985) [GR]
Kafkoula K, Papamichos N, Hastaoglou V (1990) Town plans in Greece of the 19th century. Thessaloniki [GR]
Kalflèche G (2018) Droit de l'urbanisme. Thémis Droit / Presses Universitaires de France, Paris
Kaltsonis D (2017) Constitutional history of Greece, 1821–2001. ΚΨΜ, Athens [GR]

[5] References in Greek are marked with the initials GR

Kampouroglou DG (1889) History of the Athenians: Turkish rule—First period, 1458–1687, vol. I. Alex. Papageorgiou, Athens [GR]
Kampouroglou DG (1969) History of the Athenians. 3 volumes. Palmos – N. Antonopoulos, Athens (new edition) [GR]
Karidis DN (1981) Town planning characteristics of Athens under Turkish rule. Doctoral thesis. National Technical University of Athens [GR]
Karidis DN (2014) Athens from 1456 to 1920. Archaeopress Publishing, Summertown, Oxford (UK)
Kostis K (2018) "The spoiled children of history": The making of the Modern Greek state, 18th–20th centuries. Patakis, Athens [GR]
Kotea M (1999) The constitution of Greek cities, 1828–1862. Working paper No. 33. Panteion University of Social and Political Sciences, Athens [GR]
Ladd B (1990) Urban planning and civic order in Germany: 1860–1914. Harvard University Press, Cambridge (Mass)
Lavedan P (1952) Histoire de l'urbanisme: Époque contemporaine. Henri Laurens, Paris
Lavedan P (1959) Histoire de l'urbanisme: Renaissance et temps modernes, 2nd edn. Henri Laurens, Paris
Logan Th-H (1976) The Americanization of German zoning. J Am Inst Plan 42(4):377–385
Loukatos SD (1978) The reconstruction of ruined cities in free Greece under I. Capodistria. In: Ministry of National Education and Religious Affairs (1978) Year Capodistria—200 years from his birth: Official speeches. National Printing Press, Athens, pp 79–97 and annex, pp 99–199 [GR]
Marias KG, Pantelis AM (1996) Constitutional texts, Greek and foreign. 3rd edn., Ant. N. Sakkoulas, Athens [GR]
Maurer GL von (2021) The Greek people. 2 volumes. Greek translation. Brainfood Publishing, Athens (German original 1835) [GR]
Merlin P (1994) La croissance urbaine. Presses Universitaires de France - Collection Que Sais-je?, Paris
Moskof K (1972) National and social conscience in Greece,1830–1909: The ideology of compradorist space. Thessaloniki [GR]
Nakos GP (1991) History of Greek and Roman law. University Studio Press, Thessaloniki [GR]
Olson R (2010) The Ottoman-French treaty of 1740: a year to be remembered? In: Imperial meanderings and republican by-ways <degruyter.com>. Gorgias Press. https://doi.org/10.31826/9781463231859-011
Petropoulos IA (1997) Politics and statecraft in the kingdom of Greece, 1833–1843. Greek translation. National Bank Cultural Foundation. Athens: (original in English 1968; see JA Petropulos) [GR]
Petropulos JA (1968) Politics and statecraft in the kingdom of Greece, 1833–1843. Princeton University Press, Princeton (NJ) (see also Petropoulos)
Phalip L (2019) Histoire de l'urbanisme. Bréal, Paris
Ploumidis SG (2014) The vision of Ioannis Capodistria for the Greek nation and society. In: Georgis G (ed) (2014) Governor Ioannis Capodistria: critical approaches and confirmations. Kastaniotis, Athens, pp 68–87 [GR]
Ragon M (1986) Histoire de l'architecture et de l'urbanisme modernes, vol. 1. Idéologies et pionniers 1800–1910. 3 volumes. Casterman, Paris
Rossillion P (1963) Les plans d'urbanisme. Berger-Levrault, Paris
Sakellaropoulos ThD (1991) Institutional transformation and economic development: state and economy in Greece, 1830–1922. Exantas, Athens [GR]
Schröteler-von Brandt H (2018) The history of urban planning. Academy for Territorial Development in the Leibniz Association (ARL). https://nbn-resolving.org/urn:nbn:de:0156-55997441
Steenbergen C, Reh W (1996) Architecture and landscape. Prestel, Munich
Stoye J (1988) Europe unfolding 1648–1688, 1st edn., 1969. Fontana, London

Themeli-Katifori D (1985) The French interest in Greece during Capodistria's period, 1828–1831. Epikairotita, Athens [GR]

Vakalopoulos KA (2005) History of Greece/Single volume synthesis: from the genesis of new Hellenism (1204) to modern Greece. Ant. Stamoulis, Thessaloniki [GR]

Vasiliev AA (1954) History of the Byzantine Empire, pp 324–1453. Greek translation. Bergadis, Athens (Russian original 1917) [GR]

Veremis ThM, Koliopoulos GS, Michailidis ID (2018) 1821: The creation of a nation-state. Metaichmio, Athens [GR]

Vournas T (1974) History of modern Greece: from the revolution of 1821 to the revolt of Goudi of 1909. Tolidis Bros, Athens [GR]

Wassenhoven L (2019) The ancestry of regional spatial planning: a planner's look at history. Springer, Cham (Switzerland)

Chapter 4
Governor Capodistria: The Count from Corfu

Abstract It is in the short Capodistrian period that the foundations of Greek town planning are laid. This is the subject of Chap. 4, the purpose of which is to describe the settlement policy of Governor Capodistria, the town plans of the period, and the profile of the planners, Greek or not. The author's narrative is placed in the context of the international environment and of internal divisions which mirror foreign-country antagonisms to control the character of the new state. It is significant that Greek political factions are labelled English, French and Russian. Planning affairs, as in all chapters, are discussed in the context of the international environment, domestic divisions, and local claims, regarding, e.g., the fate of the national lands and settlement creation. The chapter ends with the anomalous period after the assassination of Capodistria by disgruntled local chieftains. Shortly afterwards begins the rule of the Bavarians, following the agreement of the Great Powers to select Prince Otto as the future king of Greece.

Keywords Governor Capodistria · International environment · Domestic divisions · Town plans

4.1 Introductory Note

The main aim of this chapter is to present the continued elaborate negotiations of the European Allied Powers to determine the future of Greece, as well as the policies of Governor Ioannis Capodistria regarding the reconstruction of existing towns and the design of new settlements. Matters which are also reviewed include institutional issues, municipal governance, the contribution of foreign and Greek architects, engineers, physicians and sanitation experts, support to small cultivators, unceasing conflicts between local notables and foreign-born politicians, and, of course, the open question of the future of national lands. Although the Capodistria period is very short, an effort is made to place all these issues in a national and international context and present town planning innovations in Greece and other countries. During this phase of affairs, new problems emerge which will remain significant in the following decades.

L. C. Wassenhoven, *The History of Town Planning in a Dependent State*,
https://doi.org/10.1007/978-3-031-96958-4_4

Planning ideas are tested which are the prelude of town planning during the Bavarian period and throughout the 19th century. Information was systematically obtained by the author from archival sources, especially the General Gazette of Greece. The chapter ends with the short chaotic period after the assassination of Capodistria and before the arrival of the Bavarian Regency Triumvirate.

4.2 The International Context and the Importance of the Capodistrian Period for Town Planning

While the activities of the government of Count Capodistria are in full swing, the interventions of the Great Powers of the day are intensified. The antagonism of Britain and France is more than evident. A French military mission appeared in the Peloponnese, commanded by the vice-general Marquis Nicolas-Joseph Maison, causing intense English displeasure. In a speech, the King of France explains that the French units are in Greece only to secure peace. The British King states that the French forces will soon leave Greece, a departure confirmed by Maison. Soon, however, the commander of a French garrison in Methoni insists on its continued presence. A French scientific committee arrives in the country, made up of natural scientists, archaeologists, and architects. The Protecting Powers continue their backstage consultations and remain in touch with the Sultan's Sublime Porte, to secure the rights of Turkish subjects abandoning Greece. It is soon disclosed that discussions are under way regarding the future of wakfs, inalienable charitable endowments under Islamic Law, and of other Turkish properties. The fortresses of Athens and Vonitsa are being evacuated by Turkish forces.

Secret contacts are also taking place to select a future king, with the preferable choice being the German Prince Leopold of Saxe-Coburg, who refuses the offer and later becomes sovereign of the new kingdom of Belgium. One wonders how Governor Capodistria found time to deal with the problems of Greece, even with town planning, when the country's very existence was being gambled by the Great Powers. The diplomats of the latter turn to another candidate, given that the princes of their own royal families are excluded, although the Greeks favoured the selection of a member of the family of Queen Victoria. It is in this way that attention is turned to the royal family of Bavaria, known for its philhellenic sentiments. Soon the choice of Prince Otto is confirmed and the lay Greeks learn about this decision from a publication in a Malta newspaper. Following two interim international treaties, the final London Protocol of 1832 sealed for good the independence of the new state. It was in fact decided earlier with the Treaty of Adrianople, after a Russo-Turkish war, which also settled the status of Turkish and Greek land properties, in Greece and Turkey respectively, and confirmed the right of departing Turks to sell their properties in Attica and Euboea to the highest bidder, a development which proved decisive for the future planning of Athens. The capital of Greece was left without state-owned land available for public facilities.

The period of the Capodistria government is of fundamental importance for the history of town planning in Greece, because during its short duration key issues of planning are addressed for the first time. Several problems are tackled which have direct or indirect effect on the rehabilitation of existing cities or the foundation of new towns. It is not merely the physical form of town plans which is put to the test; it is also the social, economic and administrative side which should pre-exist in order to implement a centrally-guided style of planning, in a country without any experience of such a rational approach. The proper culture is missing and, on the contrary, the forces resisting any innovation are very strong. The resistance which Capodistria had to overcome is intertwined with interests, which one should not superficially condemn without first understanding them. For the impoverished and illiterate inhabitants of a ruined country, at least that part of it which is included in the new state, all policies originating from the top echelons of society are by definition to be feared. The potentates of Greek society, accustomed to an organizational model of autonomous power, have every reason to undermine imported values. The newly arrived educated individuals, all sorts of foreigners, who are carriers of modernizing ideas, are considered as suspect. In many cases, the motives of both sides are the expression of narrow personal interests. Intense conflicts continue to predominate, as they did during the revolution.

Regardless of the sincerity or not of their intentions, the few officials surrounding the Governor perceived the local situation in a way vastly different from the perceptions of the locals, whether those enjoying privileges, or those near starvation. Their life experiences, not excluding those of Capodistria himself, were a far cry from the expectations of the Greek peasant and soldier. The models they espoused were copied from the political and governmental systems of advanced European states, frequently authoritarian and antidemocratic, or embraced visions of cities and settlements, created under radically different circumstances.

As I mentioned before, the country's population had been decimated during the revolution. The rural economy had received mortal blows, although there were agricultural activities worthy of immediate assistance. The threats to public hygiene were severe and the danger of infectious diseases was constantly present. Entrepreneurial urban middle classes, so evident in West European cities, were virtually non-existent, some islands and port-cities being a modest exception. Trade was exposed to robbery and piracy, which demanded urgent legislative innovation. The administration itself, central or especially local, was in a state of disintegration, a situation which made necessary the urgent introduction of a governmental hierarchy.

All these deficient super- and infra-structures of the state, the disorganized economy, and a social life in disarray, were sitting on a geographical space which was in need of regulation and planning. Precedence ought to be given to existing "urban" centres, all showing signs of degradation, economic decline and insanitary conditions. Some of them deserved priority treatment because of their key-role in the national economy, trade and administration. There was also a pressing need to provide shelter to refugees from totally destroyed islands (Psara, Chios, Kassos) and unliberated regions who were seeking a place for settlement. A policy of settlement and "habitation" had to be designed as a national and patriotic priority. In this setting, requests are also made to create colonies for foreigners or Greeks of the diaspora, on the assumption that they would transfer to Greece much-needed know-how in agriculture and handicrafts. Finally, there was also a need to rebuild fortified towns which had been destroyed.

The necessary technical expertise for such an urban policy was not available locally. Trained personnel had been in demand for road construction, topographical mapping, building repair or erection, and town planning, but also health care, economic regeneration, and the revival of agriculture. All these tasks were linked with the functioning of a properly staffed administration and the associated legislative work. Trained engineers, lawyers and physicians of various specialties had to be found either through personal contacts, through the assistance of foreign military missions, or by using the services of various individuals who had arrived in the county for a variety of reasons. Two centuries later, it is of course very difficult to measure the quality of knowledge of all these experts. It is however certain that they brought with them experiences from their country of origin or education.

4.3 Domestic Developments, Towns and Town Planning

There is a variety of developments during the Capodistria period which are not limited to town planning initiatives, but have a side-effect on the regulation of space, such as management or state finances, taxation policy, military issues, trade, agriculture or education. In January 1828, the General Gazette of Greece announced the arrival on board of a British warship of the new Governor, who had earlier resigned from the service of the Russian Tzar (Laskaris ed. 1971, 13; Dafnis 2018, 553). The provisional capital was Aegina, where he remained for two years, before settling in the new capital of Nafplion. In Aegina he soon inaugurated an orphanage and a prison. His friend Jean-Gabriel Eynard, a Franco-Swiss banker and philhellene, financed the creation of a school, a teacher-training college, a hospital, and government headquarters. He later created a first bank and constructed a building in Athens, where the mint was installed. Capodistria was in regular correspondence with the Swiss benefactor (see Fig. 4.1).

The range of economic and social problems Capodistria had to address is summarized by Kotea (1999). They range from the appearance of plague and the threat of piracy, to refugee settlement and the organization of the administration. A new temporary parliament and a ministerial council were created. Spyridon Tricoupis was appointed as secretary of the interior. Special decrees were issued on financial affairs and the provincial and municipal administration. Reference to these initiatives is made in the book written much later by Von Maurer, professor of law and member of the Bavarian Regency (Maurer 2021, vol. I, 341. 344). Other decrees regulated the system of the courts of justice, the penal and commercial legal codes (GGE, vol. X, 193–194). The Governor also ordered a pioneering population and economic activity census, the organization of postal services, the revision of customs charges and the mapping of national lands. He extended his actions to the duties of

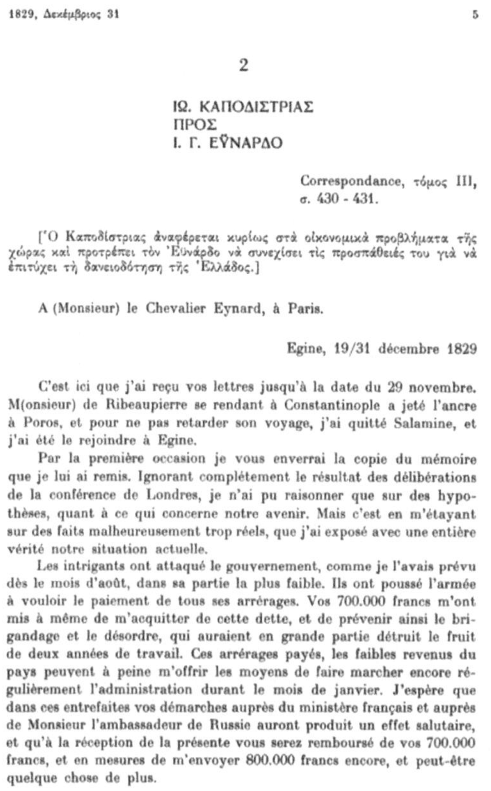

1829, Δεκέμβριος 31 5

2

ΙΩ. ΚΑΠΟΔΙΣΤΡΙΑΣ
ΠΡΟΣ
Ι. Γ. ΕΫΝΑΡΔΟ

Correspondance, τόμος III, σ. 430 - 431.

['Ο Καποδίστριας ἀναφέρεται κυρίως στὰ οἰκονομικὰ προβλήματα τῆς χώρας καὶ προτρέπει τὸν 'Εϋνάρδο νὰ συνεχίσει τὶς προσπάθειές του γιὰ νὰ ἐπιτύχει τὴ δανειοδότηση τῆς 'Ελλάδος.]

A (Monsieur) le Chevalier Eynard, à Paris.

Egine, 19/31 décembre 1829

C'est ici que j'ai reçu vos lettres jusqu'à la date du 29 novembre. M(onsieur) de Ribeaupierre se rendant à Constantinople a jeté l'ancre à Poros, et pour ne pas retarder son voyage, j'ai quitté Salamine, et j'ai été le rejoindre à Egine.

Par la première occasion je vous enverrai la copie du mémoire que je lui ai remis. Ignorant complétement le résultat des délibérations de la conférence de Londres, je n'ai pu raisonner que sur des hypothèses, quant à ce qui concerne notre avenir. Mais c'est en m'étayant sur des faits malheureusement trop réels, que j'ai exposé avec une entière vérité notre situation actuelle.

Les intrigants ont attaqué le gouvernement, comme je l'avais prévu dès le mois d'août, dans sa partie la plus faible. Ils ont poussé l'armée à vouloir le paiement de tous ses arrérages. Vos 700.000 francs m'ont mis à même de m'acquitter de cette dette, et de prévenir ainsi le brigandage et le désordre, qui auraient en grande partie détruit le fruit de deux années de travail. Ces arrérages payés, les faibles revenus du pays peuvent à peine m'offrir les moyens de faire marcher encore régulièrement l'administration durant le mois de janvier. J'espère que dans ces entrefaites vos démarches auprès du ministère français et auprès de Monsieur l'ambassadeur de Russie auront produit un effet salutaire, et qu'à la réception de la présente vous serez remboursé de vos 700.000 francs, et en mesures de m'envoyer 800.000 francs encore, et peut-être quelque chose de plus.

Source: Digital Archive of Capodistria's documents, vol. X.

Fig. 4.1 Letter of Capodistria to Eynard. *Source* Digital Archive of Capodistria's documents, vol. X

the police force, the issuing of passports, the control of wildfires, the encouragement of aquacultures and experimental farming, and the banking system. He tried to secure resources for the construction of roads and water reservoirs. His main preoccupation was the protection of the small land proprietor and cultivator. He strove to reconcile custom-law of Byzantine, even Roman, origin and modern law of French inspiration, especially in matters of land ownership, contracts, testaments, horizontal property, and surface or pre-emption rights (Pantazopoulos, n.d., 19). Capodistria repeatedly prohibited the illegal excavation and export of antiquities.

The Governor has been accused of being autocratic and was constantly embroiled in the internal feuding of Greek local primates, which dominated the national assemblies convened during his governance. In spite of the support of the national hero General Kolokotronis, his initiatives, e.g., on the workings of a senate, attracted the hostility of influential politicians, like Tricoupis, Mavrokordatos and Kountouriotis.

Being suspected as a Russophile, he was also opposed by English important figures in Greece, e.g., the generals Richard Church and Thomas Gordon, the historian George Finlay, and the ambassador Edward Dawkins (Veremis et al. 2018, 381–382). The German historian Mendelssohn-Bartholdy later accused the Governor for undermining the Greek parliament (Mendelssohn-Bartholdy 1876, 200–226, 361). In his view, national assembly deputies were being terrorized and blackmailed by the despotic regime.

When Capodistria took over his duties, the main cities of the country (Athens, Patras, Nafplion) were in a pitiful state, full of ruins, garbage and dirt. The deplorable conditions of Patras are described in a poem published in the General Gazette. Dorovinis (1985, 289–290) quotes a description of Nafplion with its stinking and bumpy lanes and its wooden, rotten and crumbling shacks. Kafkoula et al. (1990, 21–25, 30–39, 57) provide an introduction to the spatial planning work of the Capodistria administration—town reconstruction, colony building, refugee settlement, sanitation, road construction, land distribution—with the assistance of French military personnel. Administrative reforms take place with the creation of topographical and architectural departments and the appointment of Stamatis Kleanthis and Eduard Schaubert, fellow students in Berlin, as government architects.

The Greek military engineer of the French army Stamatis Voulgaris, known to Capodistria from his school days in Corfu, is charged with the production of a town plan of the Pronoia suburb in Nafplion and later of the Patras plan. Plans are commissioned to foreign engineers, i.e., to the Frenchmen Audoy for Methoni and Peytier for Corinth, Baron Charles de Schaumbourg for Messolonghi, and Baron Rudolf Artur de Borroczyn for Argos (Kokkou 1985, 359–360). Schaumbourg and Borroczyn, of Dutch and German origin respectively, had served in the Russian army. Borroczyn later worked in Bucharest.[1] Engineers of the French military corps under General Maison build the first carriage road between Methoni and the port of Navarino (Demodos 1976, 228). Mendelssohn-Bartholdy (1876, 54, 144, 357) is full of praise in his account of the contribution of the French army and accuses the Greek government for being pro-Russian, as proved by the role of "incompetent" Russian officers, like Schaumbourg and Rayko! Yet, several French engineers were very active in Greece in military works, architecture, and geodesy. Several decrees were issued regulating matters of construction, topographical mapping, house building, demolitions and repairs.

I shall return to the presence of foreign engineers, but it is necessary to mention first the requests for colonies of foreigners or Greeks, and settlements for migrants persecuted in their birthplaces. Several decrees addressed these questions. The proposal of the American S.G. Howe for the Washingtonia colony of poor Greek families near the Corinth Isthmus was accepted, and so was a settlement of Cretans accompanied by generous land grants. The building of a new town in Krissa, on the sea shore near Itea, was approved and a rather rudimentary plan was designed by A. Kalandros (see

[1] The Borroczyn plan. A short history. Muzeul Național, XXXV, 2023 . https://www.muzeulnationaljournal.ro/.

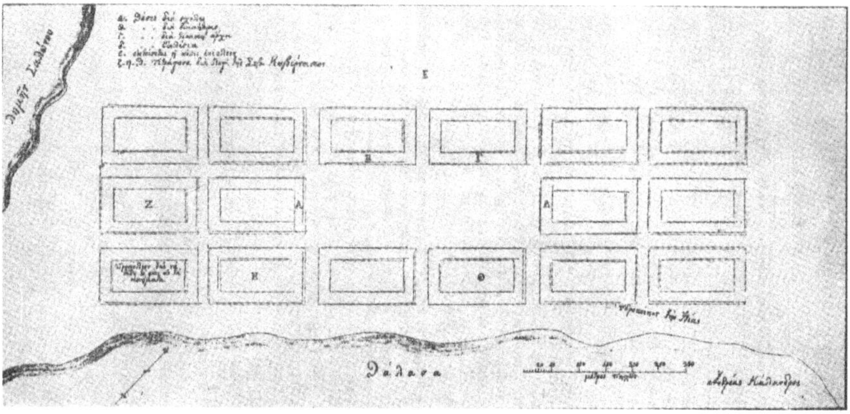

Source: Loukatos (1978).

Fig. 4.2 Plan of the new town of Krissa. *Source* Loukatos (1978)

picture 4.2). However, the proposals of the Trieste Greek community for a commercial town in Glarentza (western Peloponnese), and a similar plan for a French colony were rejected, probably because of the hesitation of the Governor to sell national lands for market ventures (Kafkoula et al. 1990, 90, 92) (Fig. 4.2).

As Loukatos (1978, 79–80), remarks, the priorities of Capodistria were determined by the obscure principle of the enlightened despotism "everything *for* the people, not *with* the people". Besides, his priority was the regeneration of existing towns. It was for this purpose that he issued the necessary orders, visited personally the damaged towns, allocated funds, and recruited a team of technocrats. In this task, he enjoyed the support of key-ministers, local builders and artisans, and simple Greek citizens who donated funds. In this context, plans were produced for a number of towns. Loukatos (op.cit., 80–95) summarized the main features of planning efforts as follows:

Main goals:

- Orthogonal planning grid;
- Division into clear neighbourhood units;
- Simple street layout;
- Clear design of entry and exit roads;
- Ample public breathing spaces, squares and commons;
- Location of administrative services, law courts, places of worship, commercial and industrial zones, education and cultural facilities, markets of agricultural products, military barracks, and places to honour the war heroes;
- Tree planting.

Additional objectives:

- Suburbs in the vicinity of large towns;

- Port facilities near important towns;
- Water supply;
- Sanitation and cleanliness;
- Architectural typology of private houses;
- Commercial and manufacturing activities and clear zoning.

Further principles:

- Reconstruction of ancient and historical towns;
- Protection of old urban quarters;
- Preservation of architectural building elements;
- Preservation of historic buildings and castles;
- Adoption of neoclassical architectural style.

It is in this spirit that the government issued many orders for the planning of towns, like Patras, Argos, Galaxeidi, and of course Nafplion, capital of the country. Land was made available for public use, residents were compensated for reasons of expropriation of their plots, and facilities were provided for the sale or continued use of existing buildings, e.g., in Nafplion, which was equipped for the first time with a public carriage link with Argos. A further worrying problem, which a government order tried to face, was the problem of diseases, especially in areas near swamps or in islands, like Lefkada in the Ionian Islands which was not yet included in the national territory (Sideris 1985, 231). It is in this period that French and English researchers published studies on insanitary living conditions.

The Capodistrian period coincided with the French efforts to secure footholds in the Eastern Mediterranean Sea to balance the British presence in the Ionian Sea and the Russian attempts to find exits to the Mediterranean and the Middle East. This of course is the background of French proposals for new towns (Themeli-Katifori 1985, 7–8, 94–106; see also Katifori 1985), which Capodistria was not inclined to accept. An officer of the French mission described him as a Russian diplomat, enemy of France and jealous of its influence in Greece. The French expeditionary force of 14,000 men under General Maison, which disembarks in Moreas in 1828, as a result of the London Protocol of 1827 (Vournas 1974, 221), includes many engineers and doctors. The French Lieutenant Colonel Garnot and the Greek-born Stamatis Voulgaris are soon involved in the design of town plans, such as that of Tripolis, a typical example of French influence and of the contrast between neoclassical and Ottoman models (Tsakopoulos 1985, 297–298, 313). Shortly afterwards, Garnot was appointed head of the new architecture and fortifications department of the government (Kokkou 1985, 360). Kardamitsi-Adami (1988) provides details on this department, the foreign experts active in Greece, and Capodistria's priorities. The Governor's friend Eynard acted as a mediator, and had asked the French Count Loverdos, a compatriot of Capodistria who worked in the French War Deparment, to recruit architects, engineers, artisans, and builders who could train Greeks for the reconstruction of towns, road-building and the restoration of monuments. The French government seized the opportunity to provide personnel who would also act as observers of the activities of the Greek administration. Present in Greece were also former officers of the Russian army, like Th. Valianos, born in Greek Cephallonia, and Baron de Schaumbourg, of Flemish descent. The Greek engineers recruited to work for the government had been educated in Russia, France, Austria and Italy and were naturally carriers of the experience of these countries. This remark applies to the Governor himself, who as a minister of the Tzar, had worked for long in Saint Petersburg, founded by Peter the Great a century earlier after his victory over the Swedes (Roberts 1996, 269). Like the Pharaoh Akhenaten and the rulers of Assyria, he had taken the bold steps of abandoning his Moscow capital to found a new capital on swampy land crossed by rivers, for geo-strategic reasons, to open a gate to Europe, as Great Alexander had done in Alexandria (Lavedan 1959, 266–273). The Russians were due to make a similar move a century later when they conquered Finland and moved its capital from Turcu to Helsinki (Wassenhoven 2019, 58). For Saint Petersburg a plan was produced in 1716 by the Italo-Swiss architect Domenico Trezzini – Trezzi according to Lavedan—who had for long worked in Holland. These must have been memories that Capodistria kept in his mind, probably also from Moscow, even more so from Odessa, founded in 1794 by the Empress Catherine the Great and designed by the Flemish engineer Franz de Voland. Odessa was one of the few cities of the Russian Empire which was built on the basis of a prior town plan.

The exact transmission of planning ideas is difficult to trace, but the fact remains that European town planning ideas travelled to Greece and brought a combination or pragmatism and neoclassicism (Kafkoula et al., 1990, 112). As stressed by Loukos (2023, 249) in his long biography of Capodistria, the surviving town plans of the period are specimens of a spirit of innovation. Sadly, Loukos reserves only a few lines to the town planning activity of the Governor.

The personality of Stamatis Voulgaris, considered to be the first town planner of modern Greece, surfaces again and again, mainly because of his 1829 plan for Patras, probably the first town plan in the sense still surviving today. Capodistria issued repeated orders for the implementation of the plan, which would involve extensive exchanges of land properties and continued to be applicable until the 1850s. Several planning historians have written about Voulgaris with great respect and admiration (Kardamitsi-Adami 1988; Dorovinis 1985. 291–294; Kyriazis 1976, 156–158). Kyriazis mentions the correspondence between Voulgaris and the famous French painter David.

4.4 The Town Plans of Patras and Corinth

The plans of Voulgaris for Patras and the Frenchman Peytier for Corinth are of particular interest not so much for their design, but rather because of the existence of the detailed letters of the planners to Capodistria, and of the correspondence on the implementation of the Patras plan (Dafnis 1976, and 1978, 69–75; Loukatos 1978, annex). With a dose of exaggeration, K. Dafnis describes Voulgaris as the first town planner of his epoch and the only town planner in the Mediterranean. He was, he argues, building new towns founded on ruins, in Nafplion, Tripolis and Patras.

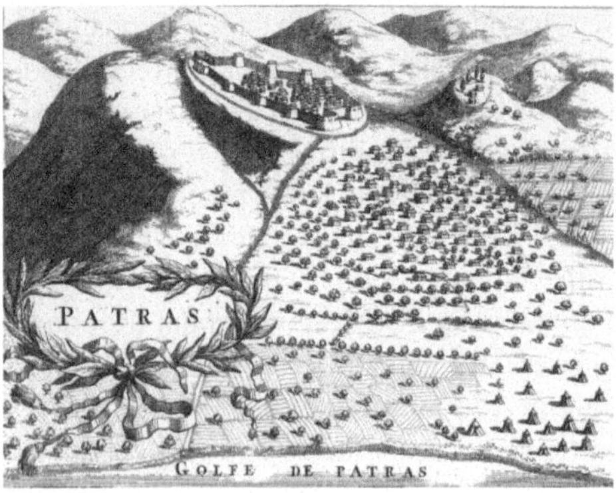

Source: Engraving by M.V. Coronelli, 1686. Lemma on the townscape of Patras in the 19th century <www.archaiologia.gr>. Free stock.

Fig. 4.3 Patras in 1686. *Source* Engraving by M.V. Coronelli, 1686. Lemma on the townscape of Patras in the 19th century < www.archaiologia.gr >. Free stock

For me, the correspondence on the plans of Patras and Corinth is significant because in it we discover concepts and procedures appearing for the first time in the history of town planning in Greece.

In the letter of Capodistria to Voulgaris of October 1838, as reproduced by K. Dafnis (1976, 53), the Governor commissions his planner to design Patras, after his successful plan for Tripolis. He points out that Turks and Arabs have already abandoned Patras, while the Greeks who return to the destroyed city ask the government to send engineers to draw the street network and the building blocks, so as to enable them to build their shacks (Fig. 4.3).[2]

In his letter of January 1829, Voulgaris – who signs as Sergeant Bulgari – presents his planning concept (Loukatos 1978, annex, 159–160) for a city which can blossom again. It will be a new town, divided into four parishes, where only some rare mud-shacks survive, along with four stone-built mosques. One of them and its side-buildings could accommodate a new Christian cathedral, a school and a library. Voulgaris proposed the opening of squares with 100 m.-wide frontages, central 20 m.-wide boulevards, secondary streets, 14 and 10 m.-wide, and 40 m.-wide port wharves. The entire street network would be linked with the existing old town and the provincial roads. Special care was reserved for open spaces, the remnants of monuments, parading esplanades, and tree-planting, for which Voulgaris had made his own contribution. He stresses that his plan had been already approved by the French General Schneider and requests the consent of the Governor. Let me underline the proposals of Voulgaris for a sort of urban zoning, the location of central functions, and green

[2] This letter, as well as all the letters discussed in this chapter, are reproduced in French in the annex of a paper by Loukatos (1978), and were translated in Greek by the present author. Apparently, the French original manuscripts were retyped and omissions were made, as well as mistakes in the terms used.

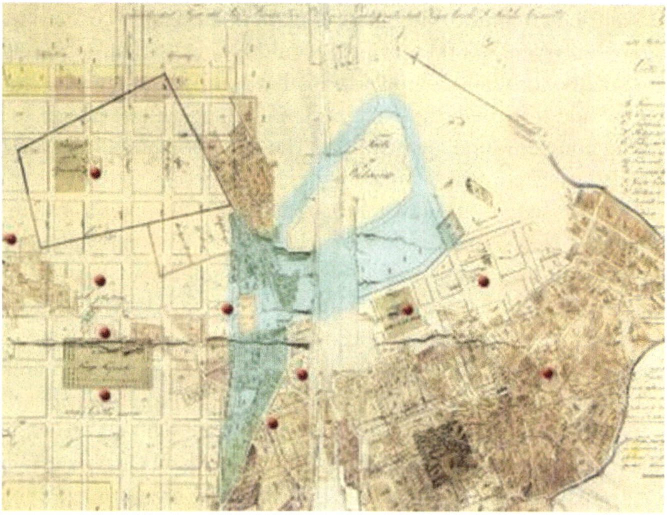

Source: Archive of Municipality of Patras.

Fig. 4.4 Plan of Stamatios Voulgaris for Patras. *Source* Archive of Municipality of Patras

spaces. In his reply, Capodistria shows a sense of realism and draws attention to the excessive width of thoroughfares, unsuitable in his view for the Greek hot and sunny climate. He also stresses his readiness to make a money contribution for the planting of trees, especially around the new town (Fig. 4.4).

A committee under the chairmanship of Baron de Schaumbourg was formed in 1830, charged with the duty to implement the plan. A long letter was addressed to the Governor in September 1830, signed by the baron, the Russian colonel Rayko, and the Greek local administrator A. Axiotis (Loukatos 1978, annex, 178–184). The French involvement is made clear in a reference to the Seine Prefecture, which sent two assistants to help the committee. Judging by their names they must have been of Greek origin. After debating the Voulgaris plan, the committee decided to seek the advice of the Greek government Compensations' Committee, in the hope that it would provide the necessary documents to determine the compensations that landowners would be entitled to receive, as a result of the implementation of the plan. The results of this consultation were disappointing and the signatories of the letter to the Governor leave a hint of suspicious behaviour, private interests and abuse.

The most interesting part of the letter concerns a decision to convene a public assembly of the Patras citizens to obtain their opinion on the future development of the city and assess their objections. The plan was presented and the details were explained, especially with regard to the previously followed procedure of distributing parcels for house construction, the building deficiencies, and the inherent planning contradictions. The committee asked the citizens to give sincere replies.

In their letter, the committee members explain that the area covered by the plan is so enormous that even 80,000 residents would be insufficient to develop it. The already sanctioned exorbitant compensations have already exhausted the space of the planned city and it is doubtful if the remaining land will be adequate to satisfy the still pending compensation applications. The city population is in sharp decline, but land property titles are still valid through inheritance or sale. The total value of properties has increased at a time when national funds available for compensation are in short supply. Hence, the residents may be in possession of large land tracks, but lack the fluidity which is necessary for the erection of new housing units. What would probably happen is that only few houses will be built, leaving huge gaps in the urban fabric. The city will not be a cohesive whole and the maintenance of streets, water ditches and sewers will be impossible. The conclusion is that a careful review of property titles and land values must precede any planning decisions. After this process, the planning zones will have to be revised, along with the types of buildings allowed in each zone. It would be only after this revision that land will be redistributed and plot owners will make decisions on the basis of the size of their allotted plots, their available means, and the possibility of joint development with their neighbours. For those still unable to build a house, there would be a further possibility to receive land, granted by the state in the periphery of the city,

The citizen assembly, after hearing the committee's presentation, authorized its representatives to reply that although the proposals seemed fair it still insisted on its disagreements. The citizens would not accept a proposal of land redistribution which was loaded with uncertainties. They wanted a precise definition of what each one could build on his existing property. The real fear was that foreigners and speculators would descend on the city and start an uncontrollable land acquisition. The citizens would prefer an immediate certainty that they can build a simple shack in the plot they possess, without expecting an ambitious replanning of the city. They insisted on the continuation of a process of compensations.

In its letter, the committee recognized that the citizens had every reason to be diffident, but insisted that there was still an open problem of land titles of dubious validity. The citizens were unfortunately under the impression that the public coffers were full of disposable resources. They envied those who might benefit from the revival of the city and were therefore ready to invest in new business opportunities and land purchases. The citizens are blinded by illusions and end by becoming enemies of themselves. The committee therefore expects that the Governor, in his wisdom, will find a solution.

The letter, in my view, touches on a broad spectrum of problems that town planning has to solve, well beyond the mere tracing of streets and design of open spaces. The Schaumbourg committee attempted an exercise of public participation long before participatory processes became a fashionable recipe after World War II. The description they provide is also a vivid image of the conflicts between traditional values and modernistic innovations. Schaumbourg wrote further letters to Capodistria, but they were now referring to his ideas for developing plants for the extraction of building materials which would reduce the cost of building.

Of great interest are also the letters of February and March 1829 to Capodistria written by the French surveyor-engineer Eugène Peytier on the plan of Corinth (Loukatos 1978, annex, 117–119). Peytier explains in detail the street layout plan, the structure in neighbourhoods, the open spaces, and the location of churches and public buildings (schools, hospitals, military barracks, courts). He makes an extensive comment on the issue of hygiene and on the local conviction that a particular wild plant (*tithymalus*) is the reason of foul air, but attributes the smell to a nearby swamp. Local residents apparently believed that a new town in Examili, near the Corinth Isthmus, would be a healthier living place. I remind the reader that Examili was the location where the American physician Howe had proposed the erection of a new settlement. What mainly worried Peytier was the physical state of the workers hired for the opening of the street network. They were hungry and ill-paid peasants, suffering from malnutrition, fed with inadequate rations of bread. Here, Peytier touches on a rarely discussed subject, i.e., the practical difficulties of implementing a plan in an impoverished country. It appears that the Governor was quick to make new funds available for the remuneration of workers, an act for which Peytier thanks him in a second letter (Loukatos 1978, annex, 121–122). He emphasizes a positive aspect of the town plan implementation, which is that it instils hope in the mind of the poor workers, that they will be able to build a house even in cases where their parcel is cut in two by a new street. It is apparent that the plan included processes of reparcelling and land readjustment, to make certain that individual plots would remain buildable.

Peytier also examined the possibility or reuse of the small ancient harbour of Lechaion, which lay in ruins, and recommended the invitation of a specialized port-engineer. His main work, however, for which he was invited to Greece, was the geodesic mapping of Moreas in cooperation with yet another French surveyor and a Russian officer of Greek descent. The primitive conditions described by Peytier leave a sad impression of the state of the country, which lacks elementary infrastructures at a time when, in France, Peytier's native country, the first railway line is being constructed (Fig. 4.5).

Peytier was an important figure in the Greek development effort. On his return to France in 1832, he published an album of maps and drawings of great interest. As remarked by S.A. Papadopoulos in his preface to the Greek edition, the geodesical mapping of the Peloponnese had been almost complete before Peytier's departure. On his arrival in Paris, Peytier was appointed in the War Archives department, as coordinator of the production of the maps of Moreas (Peytier 1971, 17). Capodistria was indeed fortunate to have such talented assistants in his entourage.

Fig. 4.5 View of Athens in the Capodistria period. *Source* Peytier (1971). *The Peytier Album*

Source: Peytier (1971). *The Peytier Album.*

4.5 Verdicts on Capodistria and the Political Chaos After His Assassination

The writings of historians evaluating the government of Capodistria are a mixture of enthusiastic comments on his innovations and condemnations of his despotic regime. Veremis and Koliopoulos (2013, 62) point out that in a country, of which the economy was in ruins, he managed to restore order in the tax system, build the foundations of a modern state, combat banditry and piracy, and introduce a decent education apparatus. His efforts were however resisted by strong local opposition. The Capodistrian administrative system of ministries, an advisory parliament, and centralized governance, met the disapproval of Anglophiles, Francophiles and local landowners, shipowners and notables (Vakalopoulos 2005, 157–158). I have already mentioned the German 19th century historian Karl Mendelssohn-Bartholdy who does not hide his hatred of the "Russophile" Capodistria. Mendelssohn-Bartholdy had written a doctoral thesis on Capodistria and in his history devoted hundreds of pages on the Governor (Mendelssohn-Bartholdy 1876). K. Dafnis fully disagrees with the prejudice of the German historian and claims that he had totally distorted reality (Dafnis 1978, 71). The assertions regarding the alleged pro-Russian policies of Capodistria are also rejected by Dakin (1972, 61–62). Capodistria's achievements, in spite of the undermining actions of local leaders, are also praised by Veremis et al. (2018, 386–388; see also Papadopoulos 1978). Although he may be accused of populism, conservatism, and autocratic values, Loukos writes (2014, 58–59, 62–67), he sincerely pursued the emancipation of the rural peasantry and the defeat of local primates and chieftains. Unfortunately, he continues, even distinguished politicians like Mavrokordatos wrongly believed that Russia was using Capodistria to install an autocratic regime. Loukos admits that the Governor sometimes displayed an autocratic behaviour, but this did not turn him into a Russian pawn or a blind enemy of British policies, although, as a native of the Ionian Islands, he had first-hand experience of English colonial values. His knowledge of colonialist attitudes is corroborated by Laskaris (ed. 1971, 122, 165–169). His support for a centralized government could be fully justified by the fragmentation of the country and did not reduce his interest in the state of agriculture and the conditions of agricultural labourers (Ploumidis 2014, 81–87). He believed in the metamorphosis of the mass of landless peasants to a people of land owners, through the distribution of land (Laskaris ed. 1971, 113–114; McGrew 1985, 69–73). This intention caused the hostility of the English ambassador Edward James Dawkins.

Capodistria's decrees constitute an elaborate legislative corpus, for which he had to rely on foreign advisors. He pieced together, according to Petropoulos (1997, 135–144) a political party of supporters, which was joined by regional groupings, but also by elements of the Russian faction. Kakouri (2021, 261–262, 343, 365) is of the opinion that Capodistria's ultimate goal was what was later described as the Great Idea of the Nation. In the short period of his administration, she adds, he achieved much more than any later politician. But a lot had still to be done. His work was left unfinished because his scheming enemies put an end to his life.

Capodistria was assassinated in September 1831 in Nafplion by members of the Mavromichali clan of primates, who were later sentenced to death. Mendelssohn-Bartholdy (1876, 373, 424). claimed that the Governor signed his own death sentence by hiding behind the Russian flag and fighting against those who remained faithful to the Constitution (!), but also reveals that according to the Austrian ambassador Prokesch-Osten fund-raising had taken place to hire assassins. The civil strife that followed wrecked the administration of the country and new assemblies were convened (Kotea 1999; Vakalopoulos 2005, 165). Avgoustinos Capodistria, brother of the assassinated Governor, was appointed temporary head of a governing committee. A futile debate ensued on the matter of national lands, land distribution, the creation of new towns, and the justice system. A new Constitution was discussed, an almost ridiculous farce, given that everybody knew that the arrival of a new king was imminent. In fact, the choice of King Otto, a Bavarian prince, was ratified in July 1832. As to the sale of national lands, it was an action resisted by the Allied Powers which considered them as pawned to facilitate international loans. However, the government made an interesting decision to enable the purchase of a national land tract by a group of co-owners. Interestingly, in Nafplion, the national capital, where the new king arrived in January 1833, a coastal zone is subdivided into building plots, which were then auctioned. A kind of early building regulation is approved, which determines building height and floor number.

The new Bavarian administration was now faced with a country in disarray and an administrative and judicial system in shambles. One additional problem is public hygiene, at a time when in Western Europe major cities suffer from a cholera epidemic. A new period is opening, which will be marked by the introduction of imported practices in various sectors, including town planning.

4.6 Conclusion

The Capodistria period ended in a most tragic way and was followed by complete anarchy. However, it left behind a host of crucial ideas, practices and innovations, which they must be singled out although they are not narrowly related to town planning. Yet, they concern a broad range of administrative, legal, social, cultural and technical issues, which form the framework of planning activity. Perhaps the most important innovation was the reform of national, provincial and local administration, a structure at the top of which stood the headquarters of the Governor himself, who provided guidance even for the drafting of a town plan. Of great importance was also the organization of the judicial system, often faced with problems of land use and ownership. Critical facets were the questions of inheritance, transactions and joint ownership. The problems of finding the right persons for staffing the administrative apparatus were quite serious and explain the reasons of resorting to the assistance of foreign missions and adventurous individuals. Some of them proved of a high calibre, either in pure politics, or in technical matters. The government had to act not only by legislating, but also through contacts and personal negotiations.

Town planning is a good example of an activity which interacted with other fields, like the protection of archaeological sites and historic settlements, welfare policy, health care, sanitation, environmental improvement, agricultural production, and assistance to social groups with special needs, like refugees and war victims. The necessary statistical or geodesic information for all these activities was absent. Capodistria in fact attempted to initiate a population census and a cadastral survey, admittedly without success, due to the shortage of the right personnel, the weak administration and local resistance. I am tempted to discern the first steps of a regional spatial policy. Perhaps, in the back of his mind, the Governor was inspired by the reforms of Jean-Baptiste Colbert, leading minister of Louis XIV in 17th century France. The central question of the land survey Capodistria was trying to carry out was the extent of national lands, not only for reasons of land use policy, but also of the management of public finances. It would take another four decades for the national lands problem to reach an end, as I will show later. Agricultural development was linked to urban development and transport facilities, thus making the nexus of policies far more complicated. A similar point can be made in connection with health care. Let us not forget that Capodistria was initially educated in medicine and had practiced the medical profession at the beginning of his career.

As it has been explained in this chapter urban policy and town planning had two components. One was the regeneration and rebuilding of existing towns; the other involved the creation of new settlements, for a variety of settlers. Very little was ultimately achieved with new towns, but the continuous efforts have shown the importance of town planning as a horizontal activity linking a wide spectrum of public policy. It is now considered self-evident, but in early 19th century Greece it was an enormous conceptual breakthrough. It was in the planning of cities like Patras, Nafplion or Corinth that these policy loops became apparent. The problems that the planners of the day were trying to solve may now seem mundane, but in Capodistria's time the shortage of material and human resources made them almost insurmountable. The timid introduction of instruments like zoning, land adjustment, and pre-emption rights, or processes like citizen participation is impressive. To convene a citizen assembly to discuss a town plan is a task which even today is admirable. Such innovations in other countries appeared gradually, but in Greece they were attempted under extraordinary conditions and in an extremely short time span. Capodistria's rule lasted for a period of less than four years, during which the government was subjected to external pressures and internal conflicts.

References[3]

Dafnis G (2018) Ioannis Capodistria. Kaktos, Athens [GR]

Dafnis K (1976) Texts by Capodistria: 200 years from his birth. Organization of Teaching Textbook Publications, Athens [GR]

Dafnis K (1978) I. Capodistria: a great Hellene. In: Ministry of National education and religious affairs (1978) Year Capodistria—200 years from his birth: Official speeches. National Printing Press, Athens, pp 69–78 [GR]

Dakin D (1972) The unification of Greece 1770–1923. Ernest Benn, London

Demodos T (1976) The first public road in Greece. In: Kyriazis P (ed) (1976) The first Greek technical scientists of the liberation period. Technical Chamber of Greece, Athens, pp 227–234 [GR]

Dorovinis V (1985) The town plan for Nafplion in the period of Capodistria, 1828–1833: the individual case and wider problems. In: SSNH/Society for the study of new Hellenism (1985a) The neo-Hellenic city: Ottoman heritage and Greek state. Conference proceedings, Vol. I. Athens, 26–28 Sept 1984, and Ermoupolis, 29–30 Sept 1984. Athens, pp 287–296 [GR]

GGE (Great Greek Encyclopaedia) (n.d.) Phoenix—Pavlos Drandakis, Athens [GR]

Kafkoula K, Papamichos N, Hastaoglou V (1990) Town plans in Greece of the 19th century. Thessaloniki [GR]

Kakouri A (2021) 1821: the beginning which was never finished, 1st edn., 2012. Patakis, Athens [GR]

Kardamitsi-Adami M (1988) The first Greek engineers: new information on the corps of Fortification Engineers and the six engineers who manned it. Technika Chronika A', 1988, 8(4): 63–89 [GR]

Katifori D (1985) The interest of expatriates for the foundation of settlements in the Greek territory, under Capodistria. In: SSNH/Society for the study of new Hellenism (1985a) The neo-Hellenic city: Ottoman heritage and Greek state. Conference proceedings, Vol. I. Athens, 26–28 Sept 1984, and Ermoupolis, 29–30 Sept 1984. Athens, pp 273–285 [GR]

Kokkou A (1985) The urban reconstruction in the period 1828–1843: State policy and reality. In: SSNH/Society for the study of new Hellenism (1985b) The neo-Hellenic city: Ottoman heritage and Greek state. Conference proceedings, Vol. II. Athens, 26–28 Sept 1984, and Ermoupolis, 29–30 Sept 1984. Athens, pp 359–366 [GR]

Kotea M (1999) The constitution of Greek cities, 1828–1862. Working paper No. 33. Panteion University of Social and Political Sciences, Athens [GR]

Kyriazis P (ed) (1976) The first Greek technical scientists of the liberation period. Technical Chamber of Greece, Athens [GR]

Laskaris M (ed) (1971) Autobiography of Ioannis Capodistria. Greek translation. Athens, Galaxias (first edition in Greek: 1940) (French original, 1868: Aperçu de ma carrière politique depuis 1798 jusqu' á 1822) [GR]

[3] References in Greek are marked with the initials GR.

Lavedan P (1959) Histoire de l'urbanisme: Renaissance et temps modernes, 2nd edn. Henri Laurens, Paris
Loukatos SD (1978) The reconstruction of ruined cities in free Greece under I. Capodistria. In: Ministry of National education and religious affairs (1978) Year Capodistria—200 years from his birth: Official speeches. National Printing Press, Athens, pp 79–97 and annex, pp 99–199 [GR]
Loukos Ch (2014) The opposition to Governor Ioannis Capodistria: re-examination of certain approaches. In: Georgis G (ed) (2014) Governor Ioannis Capodistria: Critical approaches and confirmations. Kastaniotis, Athens, pp 54–67 [GR]
Loukos Ch (2023) Ioannis Capodistria: an attempt at a historical biography. National Bank Cultural Foundation, Athens [GR]
Maurer GL von (2021) The Greek people. 2 volumes. Greek translation. Brainfood Publishing, Athens (German original 1835) [GR]
McGrew WW (1985) Land and revolution in modern Greece 1800–1881. The Kent State University Press, Kent (Ohio)
Mendelssohn-Bartholdy K (1876) History of Greece—part II. Greek translation from German original. Andreas Koromilas, Athens and Constantinople [GR]
Pantazopoulos N (n.d.) Foreword. In: Avdi-Kalkani I (ed) (n.d.) The account by Dr Gustav Geib of the law conditions in Greece. Govostis, Athens, pp 7–24 [GR]
Papadopoulos S (1978) I Capodistria, the founder of the Greek state. In: Ministry of National education and religious affairs (1978). Year Capodistria—200 years from his birth: Official speeches. National Printing Press, Athens, pp 41–56 [GR]
Petropoulos IA (1997) Politics and statecraft in the kingdom of Greece, 1833–1843. Greek translation. National Bank Cultural Foundation. Athens: (original in English 1968; see JA Petropulos) [GR]
Peytier E (1971) The Peytier album: Free Greece and the Moreas scientific committee. St. Vallianos Collection. National Bank of Greece, Athens. Introduction in Greek (Original publication in French 1832) [GR]
Ploumidis SG (2014) The vision of Ioannis Capodistria for the Greek nation and society. In: Georgis G (ed) (2014) Governor Ioannis Capodistria: Critical approaches and confirmations. Kastaniotis, Athens, pp 68–87 [GR]
Roberts JM (1996) A history of Europe. Helicon, Oxford
Sideris N (1985) Illness as incident and discourse in urban space: the case of Lefkada (1s half of 19th century). In: SSNH/Society for the study of new Hellenism (1985a), The neo-Hellenic city: Ottoman heritage and Greek state. Conference proceedings, Vol. I. Athens, 26–28 Sept 1984, and Ermoupolis, 29–30 Sept 1984. Athens, pp 251–254 [GR]
Themeli-Katifori D (1985) The French interest in Greece during Capodistria's period, 1828–1831. Epikairotita, Athens [GR]
Tsakopoulos P (1985) Tripolis: town planning, morphological study of the transition from the Ottoman to the neo-Hellenic city. In: SSNH/Society for the study of new Hellenism (1985a), The neo-hellenic city: Ottoman heritage and Greek state. Conference proceedings. Vol. I. Athens, 26–28 Sept 1984, and Ermoupolis, 29–30 Sept 1984. Athens, pp 297–325 [GR]
Vakalopoulos KA (2005) History of Greece/Single volume synthesis: from the genesis of new Hellenism (1204) to modern Greece. Ant. Stamoulis, Thessaloniki [GR]
Veremis ThM, Koliopoulos GS (2013) Modern Greece: a history from 1821. Patakis, Athens [GR]
Veremis ThM, Koliopoulos GS, Michailidis ID (2018) 1821: the creation of a nation-state. Metaichmio, Athens [GR]
Vournas T (1974) History of modern Greece: from the revolution of 1821 to the revolt of Goudi of 1909. Tolidis Bros, Athens [GR]
Wassenhoven L (2019) The ancestry of regional spatial planning: a planner's look at history. Springer, Cham (Switzerland)

Chapter 5
Bavarian Protectorate

Abstract While Otto is still a minor, government is in the hands of a three-member Regency, headed by the anglophile von Armansperg. Under the veil of the foreign protectors, the Bavarian period is, up to 1843, full of legislative activity, new town plans, settlement creation, and administrative rearrangements. It is to these activities this Chap.5 is devoted. Influential architect-planners appear on the scene, German or trained in Germany. Ambitious plans are produced for Athens and Piraeus, while eager businessmen converge in order to acquire land especially from the departing Turks. An ill-fated proposal by a German architect was to build a new royal palace on the ancient Acropolis of Athens. In this chapter, the international political and urban planning developments are reviewed, along with planning activity in Greece. A special place is reserved to the 1835 decree on the sanitary construction of towns and townships, to the plans for Athens and Piraeus, to the path-breaking legislative work of von Maurer, professor of law and member of the Regency Triumvirate, and to the influence of foreign researchers and specialized scientists who were active in Athens. Some of them are carriers of foreign philosophical currents, e.g. Saint-Simonism. To be noted that the 1835 decree was published in the Government Gazette in both Greek and German. The period ends with a public uprising in 1843 and the approval of a constitution of minor significance. With Athens becoming more cosmopolitan, it is worth remembering that the revolt in front of the palace was led by a senior military officer who had just attended an Italian opera performance.

Keywords Bavarian rule · King Otto · Town plans · Athens · Piraeus · Legislative work · Von Maurer · Foreign planners · 1835 Planning decree

5.1 Introductory Note

From the perspective of a town planner, of central interest in this chapter is the historic 1835 legislative decree on the sanitary construction of cities and townships. However, emphasis is also placed on the framework of treaties and laws with an impact on urban policy. Of key importance are the international protocols, such as

L. C. Wassenhoven, *The History of Town Planning in a Dependent State*,
https://doi.org/10.1007/978-3-031-96958-4_5

the London Protocols of 1830—before the arrival of the new king—and 1832, and the Constantinople Agreement of 1832. This is why the chapter starts with international developments and interventions in Greek affairs. Separate sections are devoted to the issues of land distribution, the Athens and Piraeus town plans, the layout plans of towns of lesser importance, the question of hygiene, the administrative organization of the country, and the legislative work of law-professor Ludwig von Maurer, member of the Regency, while Otto was still a minor. Internal antagonisms, struggles between social factions and interest groups, and dependence on the Great Powers remain as stable parameters. Discussion also continues on the problems of urban areas, the role of politicians, lawmen, architects, surveyors, and physicians, and the effect of ideas emerging in advanced countries. Attention is drawn to instances of foreign influence, either very specific, such Prussian legislation, or much broader, such as the social and philosophical currents which started with the French Revolution, Saint-Simonist philosophy, and German traditions. As in the previous chapter, the author draws information and ideas from archival sources and historiographical publications.

5.2 A New King by Divine Right and the International Scene

In the previous chapter, I mentioned the succession of protocols signed by the Great Powers, which ratified Greek independence, drew the frontier lines of the new state and granted to the Ottoman landowners of Attica and Euboea the right to sell their properties within a fixed time limit. The turning point was the 1830 protocol, which the Sublime Port had to endorse (Veremis et al. 2018, 384; Vournas 1974, 223–224). I also mentioned the choice, in essence by the British, of the Bavarian Prince Otto as the new King of Greece; a choice shared by a Greek national assembly. In this way, Carey and Carey (1968, 85) wrote, Greece became a Bavarian protectorate, of which the existence was vouched by the Protecting Powers. The British, French and Russian ambassadors in Athens were the real sovereigns of the country, even though each of them kept trying to neutralize the influence of the others. It was also thanks to the guarantees provided by the Great Powers, that an international banking loan to Greece was contracted in 1832 (Beaton 2020, 125). Stavrianos (1958, 292) has written that as late as 1841, the British ambassador had exclaimed that a really independent Greece is an absurdity. Greece, in his view was either Russian or English, and since it should not be Russian, it must be English (see also Kordatos 1973, 40, and Koryzis 1975, 111). Yet, by that time, the Great Powers had changed their policy and were in favour of securing the integrity of the Ottoman Empire (Beaton, 2020, 125), possibly because they wished to check the influence of Austria.

Greece was now under a double influence, directly Bavarian—and vaguely German—and indirectly—but decisively—British. France and Russia had lost ground. Pantazopoulos (n.d., 7), who refers mainly to legal reforms, is of the opinion that the election by the Allies of Otto as King "by Divine Right" resulted in the effacement of French cultural influence, which in earlier periods was dominant, and its replacement by the German one. However, while the French influence was motivated by the need to remedy local legal deficiencies in the sectors of commerce and shipping, the turn to German law was unjustified and dictated by an externally imposed Bavarian Regency. This regime, Pantazopoulos argues, using the royal authority, installed, by *coup d'état* a monarchy and brought to a halt the republican evolution of the country. Protagonists of this process were the Regency members Georg Ludwig von Maurer and Carl August von Abel, to whom the Hellenophile King of Bavaria Ludwig had entrusted the administrative and judicial reform of the country. They were assisted by the young jurist, and later professor of law, Gustav Geib. Pantazopoulos' account is probably exaggerated and somewhat unfair, but the truth is that the process towards a truly democratic state had been abruptly halted. There is still the open question whether the Bavarian rule was opening the door to the Europeanization of Greece, a process which is not limited to its dependence on the Great Powers.

The preconditions of this, somewhat idealized, future path were not really present. The territory of the new state, was far from being extended beyond its northern frontier, as imposed by the international protocols and accepted by the Ottoman Sultan. Greece was cut off at the southern tip of the Balkan peninsula, without terrestrial lines of communication with the advanced countries of Western and Central Europe. The economically prosperous Greek expatriates were still not persuaded that the newly-founded state offered trade and business opportunities, with the exception of certain speculators who were anxious to invest in land. They were possibly dissuaded by the internal political situation, still seriously affected by the long-distance antagonism of the Great Powers. These conditions are eloquently described by the historians Veremis and Koliopoulos (2013, 71–78). The rich Greeks of the Diaspora were reluctant to give up the affluence of cities like Vienna, Trieste, Venice, Paris and Budapest, for the sake of a miserable life in poor Athens. Political normality in Otto's kingdom was deceptive, since political parties were not functioning and the king did not provide the necessary constitutional charter. The devotion of particular political leaders to foreign power centres was encouraging conflictual tendencies. Kolettis maintained his French connection. General Kolokotronis, A. Capodistria—brother of the late Governor—and A. Metaxas relied on Russian support. Squeezed between these political formations, was the pro-British party of Alexandros Mavrokordatos. In all these parties, Veremis and Koliopoulos add, there were intellectuals, hostile to the Catholic monarch, as well as local primates, lords and military leaders who were pressing the government for a new Constitution, not so much because of democratic values, but rather to secure their return to power.

In all fairness to the Bavarian administration, it must be admitted that initiatives were taken to deal with urgent problems, like the modernization of legislation, public hygiene, administrative reform, law enforcement, and town planning. In these respects, developments in foreign countries were most instructive and their echoes must have been heard by power-holders in Greece. While Paris and Nantes were mourning thousands of cholera victims, medical scientists like Ange Guépin were carrying out extensive research (Gohier 1965; Drian, n.d.). In England, Edwin Chadwick published his report on the sanitary condition of the labouring population (Ladd 1990, 50), and laws were voted on child labour and on municipal corporations, developments of vital importance for life in cities and town planning (Cherry 1996, 3). In Germany, specifically in Düsseldorf, new building regulations are approved (Ladd 1990, 112). After a destructive fire in the centre of Hamburg, a drainage and water supply network was designed by the English engineer William Lindley (*idem.*, 50, 54). In Paris, a peripheral zone of building prohibition is imposed, while a French law introduces the possibility of expropriation for reasons of public benefit (Kalflèche 2018, 12; Gohier, 1965).

5.3 Bavarian Activity

When King Otto disembarks in Greece as a saviour, the achievements of Capodistria had been left in tatters. The Bavarian regime was to be soon despised as autocratic and dictatorial and social unrest continued until the revolt of 1843 and Otto's eventual ousting. The 21st century average Greek citizen is most likely to speak of the "Bavarocracy" with contempt and to refer to the distance separating the regime from the interests of the lower classes. A town planner of today remembers probably only the town plan of Athens by Kleanthis and Schaubert. The despotic character of the royal regime is beyond doubt and the style of decision-making was a far cry from contemporary democratic values and quite close to the European imperial or royal systems of that period, notwithstanding the philhellenic sentiments of the king and of some of his close assistants.

A differentiation has to be made between the first few years of the Regency, before Otto comes to age, and the subsequent years, after the departure of particular officials of great acumen and understanding of Greek problems. I refer to Ludwig von Maurer, member of the Regency, his right-hand assistant Geib, and other cadres of the administration, like the ideologue d'Eichtal. Yet, what undermines public confidence is the underlying authoritarianism and the endless political games between the Great Powers and key figures, like the head of the Regency Count von Armansperg. Public displeasure is accentuated by the persecution of key revolutionary leaders and the behind-the-scenes activity of obscure actors. Diplomatic pressures continue unabated with regard to, on one hand, national dreams of liberating the territories beyond the frontiers of free Greece and, on the other, debt management and loan repayment. Such pressures are perceived by the lay citizen as mediated by the royal government.

Land continues to be an issue of central importance. The inability to distribute national lands to landless peasants is attributed to foreign demands. Simultaneously, instead of calming down, the conflict grows between *autochthons* and *heterochthons*, which was already present in the years of the insurrection. As remarked by Kalyvas (2015, 39), alongside this conflict, favouritism is rife and preferential spoils' distribution continues, while the infighting never ceases between an alliance of Bavarians and *heterochthons* and the forces of local notables and former military leaders, bringing to memory the old Ottoman conditions. Mantouvalou (1988) is of the opinion that the Bavarian regime was basing its ideological justification on the argument that Greece needed external support to escape local chaos and grow into a state in the European civilization sense of the term. The trouble was that the Bavarian regime was a heavy burden on state finances, international borrowing was out of control, taxation brought citizens to their knees, and public financial management was poor (Woodhouse 1968, 158–161). Outmigration to Turkey was on the ascent, in fact much higher than the reverse flow, and the population of Greece was declining throughout the 1830s. Beaton (2020, 141) agrees that Otto's regime failed to attract Greeks living in Turkey. On the contrary, Greeks tended to search for a new fortune in Ottoman cities, like Constantinople, Izmir and Thessaloniki. It is fair to ask the question whether Athens was still undeveloped, both as a centre of economic activity and as an urban centre. Stasinopoulos (1973, 377) makes the rather ironical comment that the new capital was not yet developed in planning terms, but enjoyed celebrations and ceremonies. The desire of newcomers for quick enrichment made fertile agricultural land a most desirable acquisition, which makes Attica, the region of the capital, a preferred target. Land within the city itself is also the object of speculation, even by government officials.

Instabilities and displeasures persevere, but at the same time decisions are made which had an impact over a long-time horizon, even after the end of the Bavarian period. They concern the administration, the justice system, even town planning. The royal decree of 1835 on the planning of towns and townships would remain in force for almost a century. This is also true in the case of decrees for the implementation of the plan of Athens and subsequent amendments or extensions, but also in the case of plans of other towns or rudimentary building regulations, until the next important planning law of 1923. This makes the Bavarian period most interesting in planning history terms, as recognized in several academic studies, which were a valuable source of information, together with the original decrees published in the Government Gazette as it first appeared in 1832.

5.4 Law Making and Administration

From the first moment, the Bavarian Regency initiated a hectic legislative effort, given that until then the legal framework was either custom-law or French codes. The key person was von Maurer, a German former judge and professor of law, who managed in a span of just 17 months to produce four legal codes, which were published in the Government Gazette in German and in Greek translation. Although von Maurer had completed his legal studies in France, his work in Greece marks a turn from French to German legal science. Maurer himself wrote later that he tried to combine French simplicity with German accuracy (GGE, vol. X, 194). Of equal importance is legislation on local government which created municipalities and ten quasi-prefectures, divided into 42 provinces. The Napoleonic model was more or less imitated (*ibid.*; see also Gallant 2017, 123; Veremis and Koliopoulos

2013, 69; Vournas 1974, 239). Athens was officially named as the seat of the Royal Court. The council of ministers was also re-organized and seven ministries were established, including those for the Interior and the Economy. The new structure has been sharply criticized for its excessive centralized character, the limited powers of the council of ministers, the ultimate omnipotence of the king, the weakness of local authorities, and the failure to acknowledge the particular character of regions and localities (Kotea 1999). Local differences were totally ignored, which may explain local revolts, like that of Mani, where a Bavarian army unit suffered a humiliating defeat. Another one in Acarnania was put down by the Scottish General Thomas Gordon (Woodhouse 1968, 160).

In 1833, public spending and financial management came under the control of a Court of Auditors, of which the director was the French economist Arthémond de Regny. In spite of the existence of a separate Economy Ministry, an office of public finances is created in the Ministry of the Interior, also headed by the French Alexandre Roujoux and Gustave d'Eichtal (Vakalopoulos 2005, 171, 175), which can only be explained by the fact that the minister Kolettis was the head of the French faction. It is however necessary to remark that d'Eichtal, a Saint-Simon follower, was later the first to propose the drainage of Lake Copais and the involvement of French engineers. The most interesting part of a 1834 decree which ratified the creation of the office is the section that specifies its duties, which included geodesy and cartography, creation of colonies, protection of archaeological sites, public health and sanitation, natural landscapes, housing census, building quality, land ownership, settlement in agricultural plains of people migrating from mountain districts, plans for a road network, and negotiations with potential settlers. It is patently obvious that it was conceived as a kind of regional and urban spatial planning department! The decree was signed by all the members of the Regency and of the council of ministers, including its president Alexandros Mavrokordatos, an eminent politician I have already mentioned. The actual operation of the new office was not however smooth, as McGrew (1985, 106–107) explains. It was sabotaged by internal Regency quarrels, especially between the chief Regent von Armansperg, an anglophile, and von Maurer. D'Eichtal, member of a French banking family, was evicted, then reinstated by the minister Kolettis, and ultimately ousted from Greece, when Kolettis was deprived of his portfolio. The whole process had a very undemocratic flavour since it depended on the internal power struggle of the Regency.

The legislative work of the Bavarian administration also affected private property and land transactions, which for centuries had depended, as already mentioned, on custom-law and the Byzantine Hexabiblos by Armenopoulos. In 1835, a royal decree extended the application of traditional law until a new civil code is approved. It therefore affected real estate conditions and litigations, hence also the implementation of town and street-layout plans. Customs were closely linked to the institution of the family, the pre-emption rights enjoyed by relatives in case of property sales, and the practice of mutually-agreed brotherhood-contracts, for co-ownership or inheritance purposes (Avdi-Kalkani, n.d., 40). In the meantime, the judicial system was being revised, new law courts were created, and legal codes were adapted, usually to French prototypes, and policing of law enforcement was strengthened. Municipal police acquired the responsibility for the supervision of production activities in town centres, and also of public roads, building erection, street cleaning, ruin removal, sanitary regulation, garbage disposition, and fire protection. It was a major step towards modern practices of policing town planning infringements.

The central figure of Bavarian legislative innovations was undoubtedly von Maurer. He has been repeatedly lauded for his work, but was also the target of vicious attacks. Curiously, he resembles the case of another major figure of the period, viz. Mavrokordatos. Kyriakopoulos (GGE, vol. XVI, 655–656), speaks of a Herculean task, but admits that the laws produced by Maurer were criticized as alien to Greek traditions and were incomprehensible for the vast majority citizens. Maurer was dismissed in July 1834, because of his differences with the Regency head von Armansperg. Pantazopoulos (n.d., 8) is very caustic in his criticism of local government legislation and blames Maurer for reducing traditional local communities to ruins. This death of the community spirit remained a nationalist slogan and can be found in the writings of Tsakonas (1970 and 1971), during the 1967–1974 period of the military dictatorship. Pantazopoulos avoids to discuss the abuses of local notables, the differences between local communities, the civil conflicts, and the role of the Great Powers in installing the Bavarian regime with the consent of local leaders. In his view, Maurer and his assistant Geib, tried to disguise their real intentions by invoking the similarities between Greek and German traditional values. A further reason for the hostility against Maurer was the trial during this period of two heroes of the revolution accused of a pro-Russian conspiracy (Beaton 2020, 123). On the whole, recent writers have expressed their admiration for the enormous legislative work produced by Maurer (Woodhouse 1991, 159; McGrew 1985, 100–101; Vournas 1974, 251). Maurer, after his dismissal, returned to Germany and wrote a voluminous treatise on Greece, which was recently translated in Greek. In their prologues, Greek historians have again expressed their appreciation for the work of Maurer (Maurer 2021, vol. I, 9–23, and vol. II, 23). Gustav Geib, Maurer's close assistant and later professor at the university of Tuebingen, wrote his own account of Greek conditions and explained the application of custom-law (Geib, n.d.). It is a fashion, he wrote, to disparage the Greeks for the state of the country, but one should not forget their misery and poverty, after long years of slavery. Geib's respect for local ways of life is underlined by Avdi-Kalkani, in her introduction to the Greek edition of his book. She also emphasizes the importance of the work by Maurer and Geib for understanding the traditional legal system, for which they gathered valuable information, an effort that had started in the days of Capodistria. The situation of the country is also described in detail in Maurer's own book (Maurer 2021, vols I and II). It is fair to recognize that apart from his work as a Regency member, he left behind a valuable account of Greek geographical space and of its problems, which the management of national lands and town planning had to face in later years, even after the Bavarian period. I mean problems of land subdivision, property regulation and exchange, land readjustment, pre-emption rights and land use. Shortly after Maurer had left Greece, another major law was enacted of unknown paternity, which cannot be credited to Maurer.

5.5 The Royal Decree on the Sanitary Construction of Towns and Townships

The historic 1835 decree has been discussed by many Greek planning historians. It was signed by the three members of the Regency, including Egid von Kobell, who had replaced von Maurer. As Kafkoula et al. (1990, 62–65) write, it gives expression to an early conception of the city as a unified organism, an approach which follows the hygienic approach of the period. According to Karadimou-Yerolympos (1995, 62), it is a unique legal document, of Bavarian provenance, which proposes a succinct urban form. It has been considered by the great French planning historian Lavedan (1952, 198), as of special historical significance. Papageorgiou-Venetas (2010, 88, 246–250) claims that the principles set by the decree are those on which the 1832 plan of Athens by Kleanthis and Schaubert was based, hence it is conceivable that the father of the decree was Eduard Schaubert (see also Monioudi-Gavala 2012, 23). For Koussidonis (2014, 428), the decree is an expression both of international perceptions of urban space regulation and of the institutions and values prevailing in Greece.

A number of clauses of the decree deserve attention, i.e., those related to:

- Proximity of urban developments with transport networks;
- New settlements and migration of mountainous populations to the plains;
- Atmospheric quality, clean air, and distance from swamps, mud-pools, flood-prone areas, and volcanoes;
- Water supply from rivers and clean wells;
- Water reservoirs;
- Climatic conditions;
- Natural protection from strong cold or hot winds;
- Appropriate land relief, enabling a favourable street network and good insolation;
- Choice of street widths, size of open spaces, and tree planting locations;
- Paving of streets and squares;
- Effluents and cesspools;
- Cemeteries, hospices, hospitals and prisons;
- Noxious handicraft workshops;
- Elementary zoning of activities;
- Public facilities, like schools, churches, town halls, or hotels;
- Safe building construction standards and appropriate materials;
- Building plan approval and construction supervision by experienced master builders;
- Responsibility of municipal authorities for policing building construction.

The quality of the decree was indeed impressive. The role of prefectures, provinces and municipalities was essential. It is worth noting that Nikas (2010, 111) has found an 1847 building permission signed by the Mayor of Athens. Later, a supplementary decree introduced the concept of a green belt around towns, as a zone for promenades and the enjoyment of nature. The identity of those who drafted the 1835 decree cannot be confirmed with certainty, but they were no doubt aware of developments regarding the hygiene of cities and the techniques of town design. A number of well-educated persons surrounded the Bavarian Regency and the Royal Court and must have had at least a certain amount of influence. Great was the role of the architects Stamatis Kleanthis and Eduard Schaubert, former fellow students, who designed the plan of Athens, later amended by Leo von Klenze. All three had been educated in Berlin, capital of Prussia, and were exposed to developments in that country. It was in Prussia that an innovative piece of legislation had appeared in 1794 (*Allgemeines Landrecht für die preußischen Staaten*), which included a building regulation and granted building rights to land owners under specified conditions (Pahl-Weber and Henckel 2008). As explained by Albers (1980, 151), this law made clear these conditions and imposed limitations in order to protect the public interest, i.e., the quality of life, the security of citizens, and the preservation of the character of the urban environment. We can be certain that the authors of the Bavarian 1835 decree were fully aware of Prussian legislation.

5.6 International Town Planning Activity

The period reviewed in this chapter is very short. Nevertheless, it was in those years that certain international events took place, which it is worth mentioning given their effect on town planning history.[1] Under the command of Captain Phillip, a British vessel with a load of convicts arrived in Australia in 1788, sent by Lord Sydney. Transportation of convicts was the answer of British authorities to the problems of criminality and poverty. Colonies in Australia were used as "a receptacle for the poor", as Molony scathingly remarked. Captain Phillip first anchored in Botany Bay and later found a sheltered bay where he founded Sydney, which eventually became the capital of the state of New South Wales. In the following years the foundation was made of Melbourne, Brisbane and Adelaide, which also became state capitals. Adelaide was meticulously planned in 1836 with a classical gridiron plan drawn by William Light.[2] Australia was now receiving deported convicts, given that the doors of North America had been closed to them after the War of Independence.

Like the Byzantine Empire four centuries earlier, China was a crumbling empire dominated by a number of colonial powers. The treaties of Nanking in 1842 opened up the so-called Treaty Ports first to British nationals and later to those of other European countries or of the United States. A series of self-governed foreign settlements

[1] This short paragraph is quoted from Wassenhoven (2019).

[2] See: Hughes (1988, 87–89, 121, 442), Clark (1986, 20), Molony (1987, 87), Brunn and Williams (1983, 173, 182-183), Choay (1969, Fig. 13), Neutze (1981, pp.16–20).

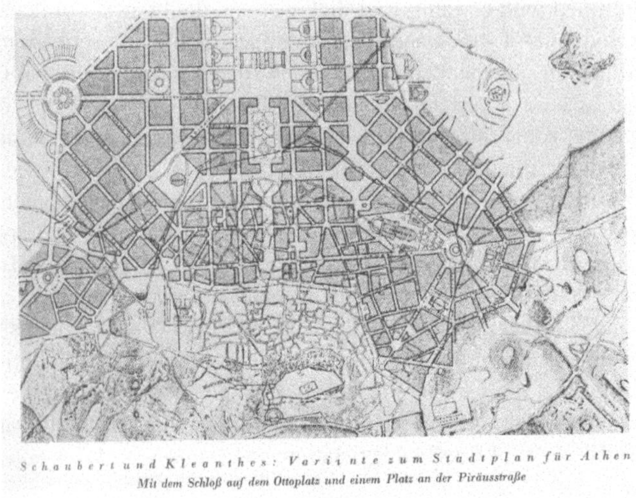

Fig. 5.1 The Athens plan of Kleanthis and Schaubert. *Source* Russack (1942). Note by the author Louis Wassenhoven: The title in German is simply due to the fact that this picture was taken from a German publication

appeared in the Treaty Ports. The parallels with the Venetian and Genoese quarters in Byzantine Constantinople are impressive. The French colonization of Algeria is also worth mentioning.[3]

5.7 Athens and Piraeus

Back in Greece, when the planning decree of 1835 was published, Athens already had its own town plan designed by the architects Kleanthis and Schaubert in 1832. It is difficult to assert that the principles of the decree influenced their work. The design itself does not lend support to such an assertion. The architects were already civil servants, but their visually attractive design rather shows that the problems of the period were mostly ignored (Fig. 5.1).

Athens, when the capital was transferred from Nafplion, was but a mere large village. The plan was in fact little concerned with the existing settlement. Kleanthis and Schaubert were rather designing a new town (Gallant, 2017, 240), which Tsiomis (1986, 20) did not hesitate to compare with Washington and Brasilia. The plan was soon replaced by a new one by Leo von Klenze, but the implementation of both was undermined by the activity of Greek and foreign speculators, who rushed to buy

[3] See: Roberts (1996, 362, 370, 377, 388), Tombs (2015 554, 574), Bergin (2015, 170, 195–197), Fredj (2017), Flitouris (2015, 137, 180, 191–192, 210–212, 217–218, 240–243, 281), Buckley Ebrey (2007, 240), Hughes (1937, 18–19, 22, 26–27, 29–30, 37–38, 41), (King 2002, 527).

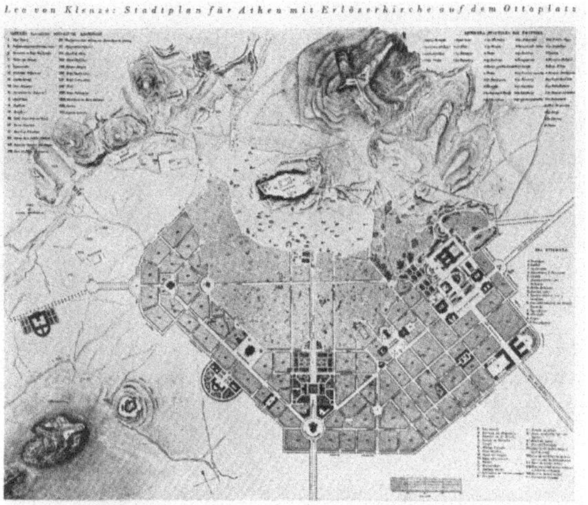

Fig. 5.2 Von Klenze's plan for Athens. *Source* Russack (1942). Note by the author Louis Wassenhoven: The title in German is simply due to the fact that the picture was taken from a German publication

land from the Turks abandoning the city.[4] A vertiginous rise of land values followed (Vournas 1974, 253). The Europeanization of the country, if indeed was taking place, was evident in the neoclassical style of the plans, which could only partially be implemented on the ground because of the resistance of land ownership interests and the shortage of public land. Still, the social and economic distance between the capital and the periphery was enormous, although, as observed by Svoronos (1972), Athens was the state capital, but the real economic capital of the Greeks was still Istanbul. Athens was certainly adorned by the royal palace, the royal gardens, new neoclassical buildings, and, above all its ancient monuments which remained untouched (Veremis and Koliopoulos 2013, 71). The most exhaustive study of the plans for Athens is that of Papageorgiou-Venetas (2010), who also produced publications on Schaubert and von Klenze (Papageorgiou-Venetas 1999 and 2000). He, too, refers to a new town that embraced the old one like a crescent, encouraged the building activity in the old town and increased the difficulty of archaeological excavations. The same author criticizes the choice of imprisoning the new town in a basin surrounded by hills, instead or developing it along an axis towards the sea coast (Papageorgiou-Venetas 2010, 62). Karidis (1990) gives us a detailed presentation of the street layout and the location of public buildings and other facilities (see also Russack, 1942). Several amendments were introduced in the plan by von Klenze (Karidis, 1990, 31–37), In his turn, Tsiomis (1985, 97–101) attempts to link the meaning of Athens as "city" to that of the "state", of which it is the symbol (Figs. 5.2 and 5.3).

It was the implementation problems of the Kleanthis-Schaubert plan that caused the decision to commission a new plan to von Klenze, although the location of the new town remained the same. Only detailed alterations were made, as explained by

[4] See: Biris (1995, 19–39; Karidis (2014, 104), Monioudi-Gavala (2015, 105–108), Monioudi-Gavala (2012, 31), GGE, vol. II, 218-221; Wassenhoven (1984, 8–10), Markezinis (1966, vol. I, 126–128), Kokkou (1985, 364), Mantouvalou (1988).

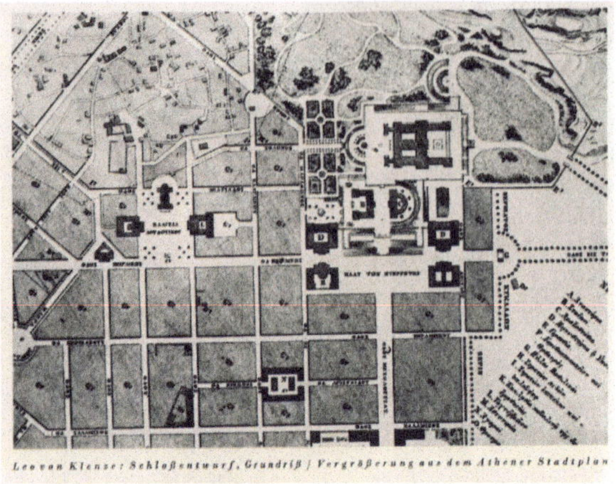

Fig. 5.3 A detail from von Klenze's plan for Athens. *Source* Russack (1942). Note by the author Louis Wassenhoven: The title in German is simply due to the fact that the picture was taken from a German publication

Papageorgiou-Venetas (2010, 129), who adds that a proposal had also been submitted, only in writing, by Alexander Ferdinand von Quast. The difficulties of implementing the original plan are also described by Maurer (2021, vol. II, 98–103), who emphasizes the violent reactions of local landowners, the intentional spreading of rumours which affected land values, and the poor adaptation of the plan to the area's topography. Maurer in fact comments that it would be wiser to postpone the transfer of the capital from Nafplion to Athens.

Whatever the deficiencies of the Kleanthis, Schaubert and von Klenze plans, as Monioudi-Gavala (2017, 19–21) remarks, these plans, along with an 1864 plan at the time of King George I, were the only statutory plans for the city as a whole during the 19th century. The numerous subsequent plans were mere local adaptations and modifications. Referring to the Kleanthis-Schaubert plan, she described it as a typical case of top-down, *ex nihilo* planning (*ibid.*, 105), Biris (1976, 59–62) draws a portrait of Kleanthis, who had participated in the insurrection led by prince Alexandros Ypsilantis in Wallachia, had studied architecture in Germany and eventually became an active architect, land investor, and businessman in Greece. In Athens he had repeated confrontations with the equally famous architect Lysandros Kaftantzoglou, not only because of isolated building projects, but also due to differences of opinion regarding the plan for Athens (Efstathiadis 1976, 172–178). Kaftantzoglou had proposed a new town between old Athens and the seafront of the Saronikos bay.

Throughout this period, a long series of decrees followed, containing the initial approvals of the plans, but also later amendments. The decrees introduced precise building regulations, penalties for illegal construction, possibilities for town plan extensions, appropriations of windfall profits, measures for the protection of archaeological sites, consultations with citizens, the provision for a peripheral green belt, and readjustment of individual land plots. The building regulations of Athens were extended to other towns, thus introducing a kind of national building regulation long before a similar stipulation in the days of King George I.

Athens had in fact become the laboratory of town planning innovations in Greece. The demolition of the Ottoman wall around the city was highly symbolical (Kostis 2018, 405). A topographical map of the city was produced in 1837 by the German officer and surveyor Ferdinand Aldenhoven, who had also published a map of the Peloponnese. His Athens map was printed in Munich and later in Athens (Aldenhoven 1841).[5] 1840 was the year of the construction of the first sewer in Athens, serving the royal palace, of which the effluents were probably discharged in an ancient stream (Hulot, 2021).

The emerging image of Athens and its distinguished architectural artefacts, designed by Stamatis Kleanthis, Lysandros Kaftantzoglou, Theofil Hansen, and Ernst Ziller have been described in detail by several historians (Vakalopoulos 2005, 235–236). McGrew (1985, 193) touches on other aspects of the growth of Athens, like the development of peripheral agricultural communities.

Of far greater importance for economic activities in Athens was the regeneration of the port of Piraeus, which had been reduced to an abandoned quay. In 1835, Kleanthis and Schaubert submitted their own town plan for Piraeus, which, as in the case of Athens, was soon modified by von Klenze (Meletopoulos 1976, 235). The entire process and the plan design are described in detail by Belavilas (2021, 49–53). Piraeus was in fact a new town. To assist its development, the Bavarian government offered incentives to refugees from the island of Chios to leave Ermoupolis and settle in Piraeus, because they were considered the most dynamic and entrepreneurial element in Greek society (Tsokopoulos 1985, 245). Free national and ecclesiastical land was offered as an incentive. A decree was issued which specified the exact area which the new settlers would inhabit. The government was putting great emphasis on developing the port city of Athens, to support commercial undertakings and export–import activity. The Athenian bourgeois class desired to have access to products, which were available in other European cities. The settlement of refugees was used as an instrument of this policy. Piraeus was not simply a promising trading port, but also a town which could accommodate industrial plants and storage facilities, which the Athenian rising middle class did not welcome in the capital. Athens would remain the centre of a Greek rich social establishment and of civil service white-collars, but Piraeus was destined to become the working-class appendage of the capital. Belavilas (2021, 163) correctly remarks that in Piraeus conditions were never the same for all social classes. Poor neighbourhoods and slum-quarters accompanied the new town from the outset. Moskof (1972, 114) adds a further background to this policy. The Protecting Powers had every reason to include Greece in the markets of their products. England and France also wished to make Greece a bastion against Russian penetration in the Eastern Mediterranean. In this context, the Greek compradors, the rich Athenian community, were a natural ally. Athens was a nest of market-oriented merchants, land owners and white-collars, served by a proletariat which, in this line of thinking, could be kept apart. At the top echelon of the social structure the dominant

[5] Internet source: TRAVELOGUES TRAVELLERS' VIEWS Places—Monuments—People/ Aikaterini Laskaridis Foundation. See also Papageorgiou-Venetas (2010, 270), who also mentions a map by August Traxel, published in Paris.

element were the expatriates (the *heterochthons*), who resided in Athens, but their economic activity was still hosted in foreign lands. The power of local primates is still considerable, but the royal entourage, which serves British interests, imposes an over-centralized decision-making process. Although the development of industry is not excluded, industry is treated as an extension of the commercial activity of the expatriate class.

Moskof's analysis, particularly when applied to the growth of Piraeus, leads to the conclusion that we witness here the convergence of—or an alliance between—a liberal ideology and a fully centralized royal regime. The deceptively liberal logic of the development of Piraeus is channelled through a hegemonic decision-making structure. Sakellaropoulos (1991, 209–211), admits that although external liberal pressures were in operation, it is doubtful whether they were successful. Rather, he argues, a compromise was attained between a variety of interests and a range of monopolistic practices were the outcome.

The colonization of Piraeus was the instrument of an unstated urban development policy, which was most certainly pursued by powers operating behind the scenes. Its outward expression was the settlement of Chios refugees, and the parallel rejection of those from the island of Psara, who were offered the alternative of Eretria, in Euboea, where they would receive land only for the erection of houses (McGrew 1985, 189). The entire process of the colonization of Piraeus and of the negotiations with the settlers from Chios and Hydra is described in detail by Belavilas (2021, 44–45, 155, 166). In due course, the city became the destination of migrants of various origins, not only islanders.

5.8 Planning Regulations and Town Plans

I do not intend to present here a full list of the town plans produced in the Bavarian period, as others have done already in Greek publications (Monioudi-Gavala 2012; Kalogiannis 2020). Karadimou-Yerolympos (1985, 383–388) has described the main ingredients of urban development policy and town building regulations:

- Land concessions to those erecting towns or suburbs, on condition that a plan has been designed;
- Availability of land plots for state or municipal services;
- Building plot size;
- Fixed deadline of building construction;
- Prohibition imposed on beneficiaries to sell their newly acquired land or erected buildings to third parties;
- Clear distinction between public and private spaces;
- Street and square alignments;
- Orthogonal arrangement of building plot facades;
- Minimum plot acreage;
- Readjustment of non-buildable plots;

- Compensation of properties traversed by streets;
- Revival of historic and ancient towns.

Kafkoula et al. (1990, 60) consider as the main features of new town plans the orthogonal grid, the division into neighbourhoods, the design of ports and wharfs, the rational street layout, and the regulation of building forms. In practice however, Karadimou-Yerolympos had remarked, the actual plan implementation proved ineffective. Most recent studies of Bavarian town planning are based on an exhaustive list of plans compiled just before World War II by Anargyros Demetrakopoulos (1934–1937, 371–372), who was later to be appointed professor of town planning at the National Technical University of Athens.

Only some of these plans will be mentioned here, but it is also essential to interpolate a brief reference to laws and decrees which determined the framework within which town planning developed. Many of these acts concerned the question of national land disposal, itself directly related to financial debts incurred by the state. National properties were distinguished between "perishable" (houses, shops, workshops, mills, olive and flour mills etc.) and "imperishable" (deserted or uncultivated land, olive groves, vineyards, currant cultivations, fish farms, salt marshes, as well precious coins and objects) (Maurer 2021, vol. I, 382). According to an 1836 decree, those national perishable properties, i.e., mainly buildings, which are not to be included in family dotation programmes, can be sold. The fate of imperishable properties, like rural wasteland, was to be decided later. The handling of national lands, in or out of towns, had an obvious impact on town plans and urban expansion, as it had been clear even in the days of Capodistria. Land was being mortgaged, not only by private owners, but also by the state itself (McGrew 1985, 66). Mortgaging was regulated by a couple of decrees of the Bavarian period.

Town plans were initially produced by engineers and surveyors, but the role of architects was on the increase, as shown in the examples of Kleanthis, Schaubert, von Klenze and Kaftantzoglou, who had been educated in other countries. To fill this gap an 1836 decree introduced an architecture Sunday school to educate builders, with the Danish architects Christian and Theofil Hansen among the teachers (Kafkoula et al. 1990, 52). A year later the School of the Arts was established, with an architectural engineering character and the Bavarian officer Friedrich von Zentner as its director. When, in 1843, von Zentner was dismissed, along with other Bavarian officials, the school switched to a more artistic direction, under its new director Lysandros Kaftantzoglou (Agriantoni 2021).

The most famous planners of the period were architects, but plans were also produced by engineers, for instance for the towns of Nafplion, Ermoupolis, and Sparta.[6] The case of Ermoupolis, port of the island of Syros, is interesting. Its plan was designed by Wilhelm von Weiler, whose house in Athens is a listed building, and is accompanied by a building regulation the application of which was extended by decree to all other towns. Further decrees on Ermoupolis, incidentally signed by

---

[6] See: Kalogiannis (2023, 161–233. 254), Demetrakopoulos (1934–1937, 370); Dorovinis (1985, 291–294), Kyriazis (1976, 163), Kardasis (1985, 585–586), Kafkoula et al. (1990, 67, 77 και 95), Loukatos (1978).

the Interior Secretary Glarakis, a Greek physician, contain a large number of details, making it, in the view of Monioudi-Gavala (2012, 24) the predecessor of the historic town planning decree of 1923.

Proposals for new townships aimed at the settlement of Germans (Maurer 2021, vol. II, 104–105) were never implemented, with the exception of the colony at Heracleion (in Attica, near Athens) for Bavarian cavalry officers (McGrew 1985, 194–195). King Otto himself gave his approval for another Bavarian settlement in Attica, but the project failed because of a lack of interested settlers. I have already referred to settlements for refugees from Chios and Hydra, for which an 1836 decree provided detailed regulations. These settlements show the extent to which town planning serves as an instrument of national policy priorities. The little-known case of the settlement in Elafonissos, an islet off the coast of Laconia, in southern Peloponnese is a very special case. It was meant to attract settlers from the mainland and offered generous land grants (Monioudi-Gavala 2012, 28). The particular interest of this settlement is that it was probably decided for foreign policy reasons, because the British were anxious to annex the islet to their Ionian Islands' possession. It is said that King Otto gave secret orders for the urgent transfer of residents and livestock.

Although it would be an exaggeration to speak of spectacular state-led new town construction, which was beyond the means of the new state, still, through methods of land grants and attractive building regulation, the outcome is not negligible. Kokkou (1985, 363) speaks of 10 new towns and reconstruction of 23 existing ones. Was there a satisfactory progress in town planning? Regarding the consolidation of a town planning mentality and the implementation of plans, the diagnosis of Demetrakopoulos (1934, 1937, 368) was negative, almost unfair. Greece, he pointed out, was not prepared for a planning reform. Poverty, lack of resources, low confidence in the state, political infighting, and other priorities made virtually impossible the radical amelioration of cities.

5.9 Land Distribution, Property Rights and Taxation

Town planning had to be carried out without violating civil law, property rights and tax legislation. Apart from the absence of cadastral records which was an obstacle in plan implementation, there was also the pressing need to increase state revenues by imposing taxes on real estate and productive activities. An 1836 decree regulated the taxation of buildings, which was imposed only on buildings which were rented to third parties and those used for a lucrative manufacturing activity. According to another decree of the same year, a tax was also imposed on activities enjoying a state-granted privilege, i.e., a preferential—perhaps monopolistic—licence. Such arrangements had to be compatible with land-use provisions of town plans.

Maintaining this delicate balance between town planning and fiscal policy required no doubt considerable sophistication of law making. It is worth recalling that in 1834 the work of Jeremy Bentham on the philosophy of law was published in Greek translation (GGE, n.d., vol. X, 194). Much closer to planning practice was the establishment of architectural and topographical departments headed by the architect Schaubert and the surveyor—"geometrician"!—D.A. Guebhard (Kokkou 1985, 362), who was charged with road and bridge construction and swamp drainage. In a decision of 1833, the Regency listed the intercity roads that had be opened, with special emphasis on linking cities with sea ports. Before returning to Bavaria, Guebhard had compiled a 180-page-long report. In spite of the provisions of an 1836 decree, a full geodesical and cadastral survey was never achieved (Vakalopoulos 2005, 235; McGrew, 1985, 131–132; Kafkoula et al. 1990, 78). It must have become clear that the Regency was in essence replicating the policies of Governor Capodistria (Kotea, 1999). Separate laws were enacted on private forests, monastic properties, land of Christianized Turks, land grants to war veterans, and the dotation of poor families. Land distribution was of course still connected with the use of national land as a guarantee for much-needed international borrowing or with its outright sale to increase public revenues. For the Bavarian government it had an additional dimension, that of enhancing its popularity among the masses. It was no accident that the family-dotation bill was written by Maximilian Frey, economics advisor of von Armansperg, head of the Regency (Gallant, 2017, 117; McGrew, 1985, 160–165). The question of agricultural land grants became even more open to criticism, when an 1838 law made them available in very favourable terms to phalangists, i.e., members of Phallanx, a veterans' association, who had been awarded initially a pension (McGrew, 1985, 177–186, 249; Ghekas, 2024, 271; Monioudi-Gavala, 2012, 36). The phalangists had no experience in land cultivation and used the certificates of land-ownership rights, which were awarded to them, in order to purchase buildings or sold them to speculators. The whole process met violent reactions by genuine agricultural peasants. The certificates became a substitute of bank notes. The veterans were of course angered by the sight of white-collars who occupied well-remunerated posts in the public administration. In the particular case of Attica, land was not available because the departing Turks had been allowed by the international independence protocols to sell their land to the highest bidder. People with special connections with the Royal Court had additional advantages. The king himself signed personal orders granting land to individuals, such as one of his attachés or a captain from Hydra, later appointed as minister of the Navy. All this turned the whole process of land grants to a charade. The centralized character of the regime was perhaps an inherent handicap.

5.10 Local Government and Health

The Bavarian government tried to gloss over its top-down system of decision-making by introducing legislation on local authorities. Reading the laws of the period reveals not only the government's intentions and the values of the period, but also the language particularities as they appear in the Greek translation of the German original. An 1834 law on the establishment of municipalities eloquently describes the Greek landscape where it relates the municipal boundaries with geographical formations, rural villages and their inhabitants, mountainous terrain, forests, and scattered dwellings. A municipality had to have at least 300 residents. A year earlier another law introduced, or, to be precise, re-introduced prefectures and provinces as a medium-tier of government. The prefects were charged with the tasks of fire protection, safety, plan implementation, sanitation, population census, road construction and maintenance, and new settlements. The laws on municipalities and prefectures were modelled on French precedents and remained valid until 1912. Decentralization may have been half-hearted but was in fact unavoidable (Kalafati, 1985, 368–369). The state, in the opinion of Kotea (2011), had to assume a transformative role, which the Greek middle class could not shoulder. Traditional communities were replaced by organized prefectures and municipalities, which were responsible for public facilities and the organization of municipal space. However, the municipal structures were uniform over all the national territory, without local differentiation and attention to the character of each place. In a sense the reform had copied the model introduced by the French National Assembly, after the French Revolution, of *départements, arrondissements*, cantons and communes. Further decrees contained details about the powers of mayors, prefects and provincial councils. The entire administrative edifice was closely related to land use planning, but was also a first sketch of regional spatial policy. It was also charged with the task of securing healthy living conditions.

Public hygiene figures highly in the agenda, because of the unhealthy situation in cities, which the 1835 town planning decree sought to remedy. Apart from the fear of diseases, concern also existed due to water shortages and the state of aqueducts and reservoirs. Kalafati (1985, 371) recalls the case of water shortages in Nafplion in 1837. An 1833 decree focused on punishable offences associated with water and food contamination, cleanliness of workshops and storage facilities, and street cleaning. The offenders were liable to be prosecuted and sentenced to imprisonment. An 1836 decree, of a strictly medical nature, dealt with the problem of infectious diseases like typhoid, cholera, plague, leprosy, or tuberculosis. Physicians occupying key positions in the administration were certainly among the authors of the decree. A leading figure was the German aristocrat Heinrich Treiber, who had received his medical education in Germany and France. He soon joined the Greek revolutionary forces and fought in the independence war. It was in his hands that General Karaiskakis died in the battle of Athens. In 1828 he was appointed director of the first military hospital in Nafplion. When Capodistria was assassinated, it was Treiber who signed the *post mortem* report and embalmed the body of the dead Governor. A series of appointments followed in key positions in the Greek army and the Royal Court, ultimately leading to a professorship in the newly created University of Athens. He was most certainly consulted when health legislation was being written. A similar observation can be made for Anton von Lindermayer, Otto's personal physician, who also opened a military hospital in Athens (GGS, vol. XVI, 156). The case of Georgios Glarakis, born in the island of Chios, is different. He was the recipient of a private scholarship and studied in Goettingen, before practicing medicine in Vienna and ultimately returning to Greece. He soon got involved in politics and served as Secretary of various ministries, i.e., Interior, Shipping, and Foreign Affairs (Veremis et al., 2018, 60; Petropoulos, 1997, 141; Vakalopoulos 2005, 185). Two more medical experts are mentioned by Tsiamis et al (2016, 55), i.e., the Bavarian physician Bernard Roeser,[7] and the pharmacist Xaver Landerer, who both served in the Royal Court. Landerer was later a professor at the University of Athens.[8]

The contacts of these medical experts with the rest of Europe, make certain their awareness of the work of famous researchers, like the French Ange Guépin and Louis René Villermé, who had published reports on contagious diseases, especially cholera, and the impact on working class housing (Gohier, 1965; Ragon, 1986, 39; Harouel, 1993, 94). They had also certainly received information about the 1830 *Treatise on Fever* of the English researcher Thomas Southwood Smith, which had been translated in French. The close association of research on hygiene and town planning is eloquently presented by Lopez (2018, 366–368), who refers to the influence of Southwood Smith's work on miasmas and "bad air". The interaction of unhygienic conditions, architecture and town planning is of course stressed in the work of Edwin Chadwick in England and then of Lemuel Shattuck, John Griscom, and Frederick Law Olmsted in the United States, which led to legislation on public health.

5.11 Conclusion

The period covered in this chapter was congested with developments, which defy any effort to summarize them and draw conclusions. I could have limited the presentation to the 1835 planning decree and the plans of Athens and Piraeus, but this would be totally inadequate. There have been events, debates and actions with serious implications for town planning. It is difficult to single out a planning parameter, which was not touched upon in the years surveyed in this chapter, regardless of the long-term practical results. A summary of concepts and notions is attempted in the following table, in alphabetical order.

[7] Author of an 1862 manual on medicine and pharmacology, with special reference to Greece (*Über die Bedeutung des heutigen Griechanland's für die Natur- und Heilkunde*. Karlsbad: Franieck).

[8] The Athens University was created with an 1836 decree and included departments of theology, medicine, law, political science and "general sciences" (philosophy, literature, natural science).

Textbox 5.1 Town planning parameters of the Ottonian period

Accessibility	Land uses
Amendments of town plans	Natural disasters
Builders' qualifications	Neighbourhood structure
Building plots	New town foundations
Building processes	Planners
Building regulations	Planning powers
Cadastral survey	Preservation of urban areas
Central city activities	Property adjustment
Climate protection	Public benefit
Colonization	Re-parcelling
Compulsory purchase	Road construction
Hygiene of cities	Settlers' accommodation
Illegal construction	Street layout plans
Infrastructures	Topographical mapping
Insanitary conditions	Town plans
Land distribution	Urban design
Land readjustment	Urban network

Apart from narrowly-defined planning parameters, this chapter, in line with the spirit of the entire book, was extended to the national and international framework, in which planning initiatives were undertaken. Hence, a number of terms appeared, listed below again in alphabetical order:

Authoritarianism	Legislation
Dependence, political and economic	Personalities
Domestic confrontations	Political factions
Europeanization	Scientific influence
Foreign prototypes	Social demands
International antagonisms	State organization
Land ownership interests	Traditional society
	University education

The importance of this chapter is due to the fact that after the ascent of King Otto to the throne and up to the constitutional reform of 1844, in spite of the omnipresent instability, the foundations were laid of the state organization for several decades. Genuine town planning initiatives also had a lasting influence, which cannot be ignored.

References[9]

Agriantoni Ch (2021) German technological superiority in Greece, 1900–1940: practices and conceptions. In: Kyrtsis, AA, Pechlivanos, M (eds) Compendium of Greek-German encounters, 1st July, 2021. URL: https://comdeg.eu/el/compendium/essay/104279/ [GR]

Albers G (1980) Town planning in Germany: change and continuity under conditions of political turbulence. In: Cherry GE (ed) (1980). Shaping an urban world. Mansell, London, pp 145–160

Aldenhoven F (1841) Itinéraire descriptif de l'Attique et du Péloponnèse, avec cartes et plans topographiques. Adolphe Nast, Rodolphe Bund et l'auteur, Athènes

Avdi-Kalkani I (ed) (n.d.) The account by Dr Gustav Geib of the law conditions in Greece. Govostis, Athens (German original by Dr Gustav Geib 1835) [GR]

Avdi-Kalkani I (n.d.) Introduction. In: Avdi-Kalkani I (ed) (n.d.) The account by Dr Gustav Geib of the law conditions in Greece. Govostis, Athen, pp 27–45 [GR]

Beaton R (2020) Greece: Biography of a modern nation. Penguin Random House, London

Belavilas N (2021) History of the city of Piraeus (19th and 20th centuries). Alexandreia, Athens [GR]

Bergin J (2015) A history of France. Palgrave, London

Biris K (1976) Stamatios Kleanthis: the grandiose idealist architect of Athens. In: Kyriazis P (ed) (1976) The first Greek technical scientists of the liberation period. Technical Chamber of Greece, Athens, pp 59–84 [GR]

Biris K (1995) Athens: from the 19th to the 20th century, 1st edn., 1966. Melissa, Athens [GR]

Brunn SD, Williams JF (1983) Cities of the world. Harper & Row, New York

Buckley Ebrey P (2007) The Cambridge illustrated history of China. Cambridge University Press, Cambridge

Carey JPC, Carey AG (1968) The web of modern Greek politics. Columbia University Press, New York

Cherry G (1996) Town planning in Britain since 1900: the rise and fall of the planning ideal. Blackwell, Oxford

Choay F (1969) The modern city: planning in the 19th century. Studio Vista, London

Clark M (1986) A short history of Australia. Penguin Books Australia, Ringwood (Victoria)

Demetrakopoulos A (1934–1937) Town plans. Techniki Epetiris tis Ellados, 1934–1937, vol. B (II), 359–449 [GR]

Dorovinis V (1985) The town plan for Nafplion in the period of Capodistria, 1828–1833: the individual case and wider problems. In: SSNH/Society for the study of new Hellenism (1985a) The neo-Hellenic city: Ottoman heritage and Greek state. Conference proceedings. Vol. I. Athens, 26–28 Sept 1984, and Ermoupolis, 29–30 Sept 1984. Athens, pp 287–296 [GR]

Drian J-C (n.d.) 1850–1995—Les étapes de la politique du logement en France. Temporalités. https://journals.openedition.org/rh19/212

Efstathiadis B (1976) Lysandros Kaftantzoglou: the first rector of the polytechnic school and its builder. In: Kyriazis P (ed) (1976) The first Greek technical scientists of the liberation period. Technical Chamber of Greece, Athens, pp 171–194 [GR]

Flitouris LA (2015) Colonial empires. Asini, Athens [GR]

Fredj C (2017) L' Algérie sera un royaume arabe. In: Boucheron P (ed) (2017) Histoire mondiale de la France. Seuil, Paris, pp 502–505

Gallant TW (2017) Modern Greece: From independence to the present, 1st edn., 2001. Bloomsbury Academic, London

Geib G (n.d.) An account of the law conditions in Greece during the Turkish rule up to the arrival of King Otto I. Greek translation (German original 1835). In: Avdi-Kalkani I (ed) (n.d.) The account by Dr Gustav Geib of the law conditions in Greece. Govostis, Athens, pp 57–170 [GR]

GGE (Great Greek Encyclopaedia) (n.d.). Phoenix – Pavlos Drandakis, Athens [GR]

[9] References in Greek are marked with the initials GR.

Ghekas S (2024) Social policy: rehabilitation of fighters, widows and orphans. In: Minakakis V (ed) (2024) The epoch of the kings: Vol. I—Otto. Kathimerini, Athens, pp 265–273 [GR]
Gohier J (1965) L'évolution de l'urbanisme en France. CRU65—Études et Essais. Centre de Recherches d'Urbanisme, Paris
Harouel J-L (1993) Histoire de l'urbanisme. Presses Universitaires de France—Collection Que Sais-je?, Paris
Hughes ER (1937) The invasion of China by the western world. Adam & Charles, London
Hughes R (1988) The fatal shore. Pan Books, London
Hulot M (2021) Underground Athens: the invisible (and real) city under the city. LIFO—Urban Culture <LiFO> [GR]
Kafkoula K, Papamichos N, Hastaoglou V (1990) Town plans in Greece of the 19th century. Thessaloniki [GR]
Kalafati H (1985) State, municipal administration and spatial organization in the 19th century. In: SSNH/Society for the study of new Hellenism (1985b) The neo-Hellenic city: Ottoman heritage and Greek state. Conference proceedings. Vol. II. Athens, 26–28 Sept 1984, and Ermoupolis, 29–30 Sept 1984. Athens, pp 367–373 [GR]
Kalflèche G (2018) Droit de l'urbanisme. Thémis Droit / Presses Universitaires de France, Paris
Kalogiannis S (2020) Town plans and public space in the Otto's period (1833–1862). Doctoral thesis, University of Ioannina, School of Fine Arts. National Archive of Doctoral Theses (www.didaktorika.gr) [GR]
Kalogiannis S (2023) Town plans and public space at the beginning of the Greek state. Nomiki Vivliothiki, Athens [GR]
Kalyvas SN (2015) Modern Greece. Oxford University Press, New York
Karadimou-Yerolympos A (1985) Planning and recovery of city space: character of state planning interventions during the transition from the Ottoman to the neo-Hellenic city. In: SSNH/Society for the study of new Hellenism (1985b) The neo-hellenic city: Ottoman heritage and Greek state. Conference proceedings. Vol. II. Athens, 26–28 Sept 1984, and Ermoupolis, 29–30 Sept 1984. Athens, pp 381–395 [GR]
Karadimou-Yerolympos A (1995) The reconstruction of Thessaloniki after the 1917 fire. University Studio Press, Thessaloniki [GR]
Kardasis V (1985) Ermoupolis: urban functions and behaviour. In: SSNH/Society for the study of new Hellenism (1985b) The neo-Hellenic city: Ottoman heritage and Greek state. Conference proceedings. Vol. II. Athens, 26–28 Sept. 1984, and Ermoupolis, 29–30 Sept 1984. Athens, pp 585–589 [GR]
Kardasis V (2014) Governor Capodistria in confrontation with the traditional social strata. In: Georgis G (ed)Governor Ioannis Capodistria: critical approaches and confirmations. Kastaniotis, Athens, pp 108–121 [GR]
Karidis DN (1990) Reading town planning: the social meaning of spatial forms. Athens [GR]
Karidis DN (2014) Athens from 1456 to 1920. Archaeopress Publishing, Summertown, Oxford (UK)
King AD (2002) Extract from "Urbanism, colonialism, and the World-economy" (Routledge, 1990). In: Bridge G, Watson S (eds) The Blackwell City reader. Blackwell, Oxford, pp 524–534
Kokkou A (1985) The urban reconstruction in the period 1828–1843: state policy and reality. In: SSNH/Society for the study of new Hellenism (1985b) The neo-Hellenic city: Ottoman heritage and Greek state. Conference proceedings. Vol. II. Athens, 26–28 Sept. 1984, and Ermoupolis, 29–30 Sept. 1984. Athens, pp 359–366 [GR]
Kordatos G (1973) The English interventions in Greece. Epikairotita, Athens [GR]
Koryzis Ch (1975) The authoritarian regime 1967–1974. Gutenberg, Athens [GR]
Kostis K (2018) "The spoiled children of history": The making of the modern Greek state, 18–20th centuries. Patakis, Athens [GR]
Kotea M (1999) The constitution of Greek cities, 1828–1862. Working paper No. 33. Panteion University of Social and Political Sciences, Athens [GR]

Kotea M (2011) The rise and fall of mayoral power in Greece, 1833–1936. Int J Humanit Soc Sci 2011(1):13–20

Koussidonis Ch (2014) Regulations of space in the new Greek state and the decree on the sanitary construction of towns and townships of 1835. In: Drakoulis DP, Tsotsos GP (eds) Historical, social and town planning analysis of space: Tribute to Prof. Evangelos Demetriadis. Antonios Stamoulis, Thessaloniki, pp 427–443 [GR]

Kyriazis P (1976) Stamatis Voulgaris: the fighter, the town planner, the man. In: Kyriazis P (ed) (1976) The first Greek technical scientists of the liberation period. Technical Chamber of Greece, Athens, pp 151–170 [GR]

Ladd B (1990) Urban planning and civic order in Germany: 1860–1914. Harvard University Press, Cambridge (Mass)

Lavedan P (1952) Histoire de l'urbanisme: Époque contemporaine. Henri Laurens, Paris

Lopez R (2018) Public health and urban planning: Intertwined histories. In: Hein C (ed) (2018) The Routledge handbook of planning history. Routledge, London, pp 366–376

Loukatos SD (1978) The reconstruction of ruined cities in free Greece under I. Capodistria. In: Ministry of National education and religious affairs (1978) Year Capodistria—200 years from his birth: official speeches. National Printing Press, Athens, pp 79–97 and annex, pp 99–199 [GR]

Mantouvalou M (1988) Town planning of Athens (1830–1940). In: Sakellaropoulos Ch (ed) (1988) From the Acropolis of Athens to the port of Piraeus: renewal plans of urban areas. National Technica University of Athens—Politechnico di Milano, Athens [GR]

Markezinis S (1966) Political history of contemporary Greece, 1828–1964. 4 volumes. Papyros, Athens [GR]

Maurer GL von (2021) The Greek people. 2 volumes. Greek translation. Brainfood Publishing, Athens (German original 1835) [GR]

McGrew WW (1985) Land and revolution in modern Greece 1800–1881. The Kent State University Press, Kent (Ohio)

Meletopoulos I (1976) The creation of Piraeus. In: Kyriazis P (ed) (1976) The first Greek technical scientists of the liberation period. Technical Chamber of Greece, Athens, pp 235–248 [GR]

Molony J (1987) History of Australia: the story of 200 years. Viking—Penguin Books Australia, Ringwood (Victoria)

Monioudi-Gavala D (2012) Town planning in the Greek state, 1833–1890. Department of Cultural Environment Management and New Technologies, University of Western Greece, Agrinion [GR]

Monioudi-Gavala D (2015) The Greek city from Hippodamus to Kleanthis. Kallipos/Open Academic Editions (www.kallipos.gr), Athens [GR]

Monioudi-Gavala D (2017) Planning and real estate property in Athens, 1833–1922. Paraskinio, Athens [GR]

Moskof K (1972) National and social conscience in Greece, 1830–1909: the ideology of compradorist space. Thessaloniki [GR]

Neutze M (1981) Urban development in Australia. George Allen & Unwin, Sydney

Nikas GA (2010) The issuance of building permissions from the formation of the modern Greek state to the publication of the Legislative Decree of 17 July 1923. Tech Chron Sci Ed 2010(1):109–119 [GR]

Pahl-Weber E, Henckel D (eds) (2008). The planning system and planning terms in Germany: A glossary. Academy for Spatial Research and Planning (ARL), Hanover

Pantazopoulos N (n.d.) Foreword. In: Avdi-Kalkani I (ed) (nd). The account by Dr Gustav Geib of the law conditions in Greece. Govostis, Athens, pp 7–24 [GR]

Papageorgiou-Venetas A (1999) Eduard Schaubert, 1804–1860: collection of evidence for the planning of Athens and Piraeus. Translation from German original. Odysseas, Athens [GR]

Papageorgiou-Venetas A (2000) Leo von Klenze in Greece: evidence for neo-Hellenic historiography. Odysseas, Athens [GR]

Papageorgiou-Venetas A (2010) Athens: A vision of classicism. Greek translation from German original. Kapon, Athens [GR]
Petropoulos IA (1997) Politics and statecraft in the kingdom of Greece, 1833–1843. Greek translation. National Bank Cultural Foundation. Athens: (original in English 1968; see JA Petropulos) [GR]
Ragon M (1986) Histoire de l'architecture et de l'urbanisme modernes. Vol. 1. Idéologies et pionniers 1800–1910, vol 3. Casterman, Paris
Roberts JM (1996) A history of Europe. Helicon, Oxford
Russack HH (1942) Deutsche Bauen in Athen. Wilhelm Limpert Verlag, Berlin
Sakellaropoulos ThD (1991) Institutional transformation and economic development: state and economy in Greece, 1830–1922. Exantas, Athens [GR]
Stasinopoulos EK (1973) History of the Athenians: from antiquity to our age. Athens [GR]
Stavrianos LS (1958) The Balkans since 1453. Holt, Rinehart & Winston, New York
Svoronos N (1972) Histoire de la Grèce moderne. Presses Universitaires de France, Paris (1ère édition 1953)
Tombs R (2015) The English and their history. Penguin, London
Tsakonas D (1970) Problems of Hellenicity: sociology and philosophy of contemporary Hellenism. Athens [GR]
Tsakonas D (1971) Introduction to new Hellenism. Athens [GR]
Tsiamis K et al (2016) The diachronic development of infectious diseases and microbiology in Greece in the 19th Century, through the scientific presentations at the Medical Society of Athens (1835–1900). Microbiology Laboratory, School of Medicine, National and Capodistrian University of Athens. Iatrika Chronika 61(1), Jan-Mar [GR]. https://doi.org/10.5281/zenodo.10020734
Tsiomis I (1985) Athens: a European affair. In: SSNH/Society for the study of new Hellenism (1985a) The neo-hellenic city: Ottoman heritage and Greek state. Conference proceedings, vol. I. Athens, 26–28 Sept 1984, and Ermoupolis, 29–30 Sept. 1984. Athens, pp 97–101 [GR]
Tsiomis I (1986) Parler d'Athènes de 1834, comme on parle de Brasilia de 1964. In: Burgel G (ed) (1986) De la Polis aux politiques urbaines: renaissance et mutations de la ville grecque contemporaine. Villes en Parallèle. No 9, p 20
Tsokopoulos V (1985) The stages of local conscience: Piraeus, 1835–1945. In: SSNH/Society for the study of new Hellenism (1985a) The neo-hellenic city: Ottoman heritage and Greek state. Conference proceedings, vol. I. Athens, 26–28 Sept 1984, and Ermoupolis, 29–30 Sept. 1984. Athens, pp 245–249 [GR]
Vakalopoulos KA (2005) History of Greece/single volume synthesis: from the genesis of new Hellenism (1204) to modern Greece. Ant. Stamoulis, Thessaloniki [GR]
Veremis ThM, Koliopoulos GS (2013) Modern Greece: a history from 1821. Patakis, Athens [GR]
Veremis ThM, Koliopoulos GS, Michailidis ID (2018) 1821: The creation of a nation-state. Metaichmio, Athens [GR]
Vournas T (1974) History of modern Greece: From the revolution of 1821 to the revolt of Goudi of 1909. Tolidis Bros, Athens [GR]
Wassenhoven L (1984) Greece. In: Wynn M (ed) Planning and urban growth in Southern Europe. Mansell, London, pp 5–36.
Wassenhoven L (2019) The ancestry of regional spatial planning: A planner's look at history. Springer, Cham (Switzerland)
Woodhouse CM (1968) The story of modern Greece. Faber & Faber, London
Woodhouse CM (1991) Modern Greece: a short history. Faber & Faber, London

Chapter 6
National Lands, Clientelism and Tottering Town Planning

Abstract The Ottonian reign ended in 1862 after a new revolt demanding a truly democratic constitution. Otto is expelled and a new king ascends to the throne, this time a Danish Prince, George I, once again selected by the Great Powers, mainly Britain, which cedes to Greece the Ionian Islands, as a gesture of good will. The annexation of the region of Thessaly follows in 1881. All these developments are included in this chapter, along with urban planning innovations in third countries, where the housing and health problems of the working classes are given a great deal of attention. In Greece, this period is marked by the long-awaited distribution of national lands to landless cultivators and the abolition of old Turkish *chifliks*, an action which results in a fragmentation of properties into small parcels. This phenomenon was to have extremely serious results for building activity in out-of-plan agricultural areas. Thessaly, with its large assets of rural land, is especially affected.

Keywords Eviction of King Otto · King George I · Territorial annexations · National land distribution · Parcelling of estates · Clientelism

6.1 Introductory Note

This chapter is mainly devoted to the period which starts with the revision of the Greek Constitution in 1844 and ends with the annexation of Thessaly in 1881. However, it was deemed essential to begin with the nineteenth century expansion of colonialism all over the globe and the associated creation of towns and settlements by the European powers. In Europe, the public interest turns to issues like housing, health, urban rejuvenation, radical—in fact, brutal—renewal of cities, as in Paris, legislative regulation of urban development, transfer of powers to municipal authorities, and infrastructures. Pioneering industrial settlements are founded.

Further, developments in Greece are reviewed, with special emphasis on the 1844 and 1864 constitutional amendments, which changed administrative organization and national land management, thus affecting the structure of national space and towns.

L. C. Wassenhoven, *The History of Town Planning in a Dependent State*,
https://doi.org/10.1007/978-3-031-96958-4_6

A separate section is devoted to the 1871 distribution of national lands, that gradually modified urbanization patterns. Political instability, particularly in the 1860s, favoured patronage and clientelism, with an indirect effect on town planning. Plans are constantly being amended with hasty procedures, Athens being a typical example, to satisfy a multitude of private interests. The effort of creating refugee settlements continues, an inheritance of the revolutionary past.

6.2 International Developments and Colonialism

Britain and France remained the main Protecting Powers of Greece, but along with other European countries entered a phase of hectic antagonism for the expansion of their colonial empires, along with other European countries, such as Portugal, Germany, Holland, and Russia.[1] In the course of colonialist expansion a range of new towns mushroomed in Africa, Asia, North and South America. In South Africa, Cape Town was first colonized by Portuguese explorers, but it was the Dutch who established a supply station and a fortified camp in mid-seventeenth century. Cape Town grew into a busy trading post, received settlers from Holland and France and was captured by the British at the end of the eighteenth century. Pretoria, founded by private initiative, became the capital of the South African Republic in 1855. Dakar, since 1960 the capital of Senegal, is the most appropriate example of a colonial city. It passed through the hands of the Portuguese, Dutch and English, before its final capture by the French, who, in 1857, turned the old settlement of Ndakaaru into a military station and changed the name to Dakar. My next example is taken from Asia. Jakarta is Indonesia's capital and largest city. It was called Jayakarta when the Dutch arrived there in 1596, but it was renamed by them Batavia when it became the capital city of the Dutch East Indies. In Latin America, Rio de Janeiro was founded by the Portuguese in 1567, on the site of a short-lived French settlement, but it was during the Napoleonic wars that it served as capital of the entire Portuguese empire.[2] The importance of this territorial consolidation can be appreciated by the immensity of the Spanish American empire by 1700, in South and North America, extending from River Plate to Colorado and from southern Chile to northern California and other parts of the present United States, e.g. Florida, as Roberts aptly remarks. Spanish urbanization was in a way "rewriting geography". As Hardoy has written, "with the exception of Montevideo, all metropolitan areas that had more than one million inhabitants in 1960 were founded in the sixteenth century". All principal cities, of which Hardoy provides a list, were in place before the end of the century. The creation of new settlements was not an unknown practice in the empires which the Spaniards destroyed; the Incas, e.g., were applying a systematic settlement policy in areas they had conquered. Apart from settlement officially sponsored by the Spanish Crown,

[1] In these paragraphs, I am quoting myself (Wassenhoven 2019).

[2] See: Williamson (2009), Pendle (1967), Prescott (1979), Brunn and Williams (1983), Flitouris (2015), Legge (1980), Monteil (1972).

a major role was played by missionaries who resettled Indians in *congregaciones*. "These model towns", according to Williamson, "were inspired by the ideal society described in the *Utopia* of Sir Thomas More". However, the policy of concentrating natives in special settlements, often by force, was also practiced by the official colonial authorities. It was again practiced in Mexico in the nineteenth century, after independence.[3]

In the race of imperial colonialism, the Portuguese were to be followed by the British, French, Spanish, Dutch and Danes. What distinguishes English settlement from, e.g., Spanish colonization, was the remarkable number of settlers who were not conquerors but were rather looking forward to a new home, and to the creation of new communities, which would prosper by land cultivation. They were also relying on private initiative, the pursuit of opportunities, and self-government. George has written that the Europeans of the colonial era, in their effort to colonize new lands and to exploit their resources, needed three categories of settlers: builders, cultivators and miners. The Spaniards imposed strict control on the immigrants admitted to their American empire, in sharp contrast to the British attitude. Puritan settlers, stresses Trevelyan, were villagers in origin and were not led by feudal nobles as in the case of French colonization in Canada. The first English settlers emigrated as entire families in the hope of finding a new farming environment. Their right to settle still required a charter, but their colonies were self-governed and not sponsored by the British government, although gradually self-government was eroded by the royal authorities. In an interesting remark, Roberts finds a parallel with ancient Greece: "Like the colonies of the ancient Greek cities, such colonies were usually soon substantially independent of the mother country".[4]

Intra-European struggles had a direct impact on the colonies. By the eighteenth century, France had become a colonial power with possessions, though not comparable to those of Britain, in New France (Canada and the Mississippi Valley), the Caribbean islands, and in trading posts in Africa and India (Senegal, Madagascar, and Pondicherry in India). The "craze" for sugar made the West Indies a field of commercial clash, which explains the value of British Jamaica and the French islands of the Antilles (Martinique, Guadaloupe, and Saint-Domingue, now Haiti, which gained its independence in 1804). The Anglo-French conflict in India was both more complex and violent. In India the British government assumed the role of the East India Company. The "state" officially replaced the company in 1858, "in name, as it had long done in fact", in Trevelyan's words. British town planning concerned mostly military cantonments and civil stations for administrators and officials. At about mid-eighteenth century the English had captured Quebec and driven the French out of Ohio, renaming Fort-Duquesne as Pittsburgh. An indirect result of the French withdrawal from North America, following the European Seven Years' War of that period, was that the first phase of the British Empire was now over. When French

[3] See: Williamson (2009), Pendle (1967), Prescott (1979), Flitouris (2015), Bernstein (2009), Hardoy (1975a, b), Rodriguez (2005), Monclús and Díez Medina (2018).

[4] See: Trevelyan (1952), Roberts (1996), Tombs (2015), Brogan (2001), Pendle (1967), George (1964).

Canada was taken by the British a flow of British colonists entered Canada, either originating in the United States, after the War of Independence, in late eighteenth century, or coming from Britain in the nineteenth century.

At the start of the nineteenth century the rise of the second British Empire, comprising India, was without precedent. There was a constant stream of emigrants from Great Britain to various parts of the empire. Migration flows included Germans settling in Argentina and Chile and Asians recruited to work in South America and the West Indies. It was what Roberts calls the Great Resettlement, involving tens of millions of Europeans, a movement towards colonies, which continued even after some of these colonies had become independent nations. In the nineteenth century the British Empire covered twenty per cent of the earth's land surface. Britain kept annexing lands in Natal, the Gold Coast (now Ghana), New Zealand and Fiji. Competition among European powers was, however, still fierce. As Lichtheim remarked, the British, French and Russian empires were soon to be challenged by Germany and Japan. The British captured Malaysia, Singapore, Hong Kong, the British Honduras (now Belize), and Aden. In the late nineteenth century, the colonies of Rhodesia and Transvaal were created. The French lost Haiti but retained some islands in the West Indies, Reunion (*Réunion*) in the Indian Ocean, Guiana in Latin America and other colonies.[5]

Africa was the great disputed prize of the nineteenth century, in a contest in which the French strove to be the protagonists. They annexed Algeria, in theory as a territory of France, which soon attracted a huge flow of immigrants and investment from France and generated vigorous trade. A somewhat haphazard acquisition of possessions (Senegal, Ivory Coast and Tahiti) followed the disastrous war of 1870–71 with Prussia. In the 1880s started what came to be known as the "scramble for Africa", initially along the continent's Mediterranean coast where the vacuum of former Ottoman territories offered opportunities. There followed the French protectorate of Tunisia and the British occupation of Egypt. France created protectorates in Indochina and Madagascar and in the early years of the twentieth century founded the West Africa and Equatorial Africa federations and acquired Morocco, in the face of competition from those who were trying to set foot in Africa, as were the Italians in Libya and Ethiopia. The king of Belgium set up a private empire in the Congo. However, opportunities for European migrant-settlers were not always favourable. Imperial Russia was also involved in the nineteenth century imperial expansion towards Siberia, central Asia, Persia, and China. In Siberia, Vladivostok was founded in 1860. Expansion took place in North America too, Alaska being the main example, ultimately sold to the United States. The main loser was China, where in the nineteenth century events were taking a traumatic turn.

[5] See: Trevelyan (1952), Tombs (2015), Lichtheim (1971), Flitouris (2015), Jenkins (2012), Roberts (1996), Money (1976).

6.3 Town Planning and Housing in Europe

Significant urbanization change was taking place in Europe itself. Perrot remarks that it was in the eighteenth century, that the urban locational pattern in France, focused as it was on a limited number of cities and ports, began to evolve towards a network including provincial capitals. It was towards the end of the century, in Pinol's view, that the definition of a town ceased to depend on the fortifications that surrounded it and came to be associated with economic functions. An irony of history is, as Ragon indicates, that the radical re-planning of Paris by the prefect Baron Haussmann in the nineteenth century had a military and policing logic. Haussmann's Paris was not of course a new city, but the extent of his interventions justifies their inclusion in the present section. To quote Peter Hall, "Haussmann was embarking on an extraordinary job, never before undertaken in human history: the rebuilding of an entire city".[6]

I do not intend to discuss the large conurbations of European industrial countries and their growth in the nineteenth century, but I have to say a few words about urban initiatives which foreshadow the new towns of the twentieth century. The nineteenth century witnessed the creation of a number of residential communities, in the form of garden suburbs, and of workers' towns around factories, built by progressive industrialists. Choay listed the most important of these early satellites, in Paris and London, and new towns created by industrialists. The nineteenth century is equally famous for the host of writings on utopian cities, e.g. by Fourier, whose ideas were put into practice by the industrialist Godin who built a workers' town (Benevolo 1967; Choay 1969).

In the urban centres of major European countries, a labour proletariat is growing, often in a state of revolt. One could claim that the anti-royal insurrection of 1843 in Greece was an early sign of unrest in Europe, although Greece did not yet have an industrial proletariat and the insurrection ended with a compromise. In France however, the 1848 uprising put an end to the royal regime of king Louis-Philippe I and installed the Second Republic (Bergin 2015). That was the year of the dismissal of Metternich, the autocratic Chancellor of Austria, well-known for his interference with Greek affairs (Vournas 1974, 305–306).

Apart from democratic demands, a serious cause of social unrest was the deplorable living conditions of the working class. A revolt of industrial weavers took place in Germany in 1844 (Ragon 1986, 32), and a year later the historic book by Friedrich Engels is published on the condition of the working class in England, a document which town planning historians never fail to mention (Ragon 1986, 32; Phalip 2019, 50; Choay 1965, 181–191; Benevolo 1967, xiv), for obvious reasons. This condition explains an array of initiatives in the fields of town planning, housing and sanitation. In 1847 and 1848, laws are enacted in England on town improvement, policing, cemeteries, markets, exhibitions, and public health. Several laws follow on working class housing (Heap 1969, 2–5).

Housing, town planning and public hygiene are the key facets of multiple government measures. In France, in 1848, the letting of insalubrious houses was forbidden

[6] See: Perrot (1992), Pinol (2000), Ragon (1986), Choay (1969), Hall (1998).

by law. A second law followed in 1850 on insanitary dwellings, which the well-known planning historian Jeanne Hugueney considers as the first French town planning act (Bourillon 2000; Drian n.d.; Hugueney 1950). As pointed out by Drian, it was certainly the beginning of a consistent housing policy. Housing conditions were blamed for the 1832 Paris cholera-epidemic, which caused 18,600 deaths. The 1850 act defined the meaning of insalubrity and ordered the municipalities to oblige house owners to carry out the necessary improvements. Although the implementation of the law left a lot to be desired, the new approach in favour of public local action to improve housing conditions was a major step.

In London, pollution was reaching a critical level. 1858 was the year of the Great Stink (Mak 2008). Geert Mak refers repeatedly to the pioneering work of Charles James Booth of the 1880s and his description of the appalling living and working conditions in London, of which Mak gives a detailed account. Pollution, sewage, stench, dirt, smog and darkness were the rule. Laws were enacted in 1868 and 1875 to improve living conditions and protect public health (Torrens' Act and Cross' Act), which also enabled compulsory purchases (Ashworth 1968, 181; Heap 1969, 2–5). Similar concerns about hygiene and housing were dominant in the 1860s in Germany, where the notorious multi-floor Rental Barracks (*Mietskaserne*) appear (Ladd 1990, 78–79, 83; Mak 2008, 43–44). Ladd also discusses the displacement of affordable housing from city centres, the speculative activity of building contractors, and the consequent responsibility of town planning. Similar conditions had appeared earlier in England, labelled in Germany as *Manchestertum* (Kafkoula 2007, 131). The growing hygienic problems of cities gave birth to public health societies, medical conferences, and the increasing involvement of physicians like Franz Varrentrapp and Carl Reclam, who emphasized the role of engineers (Ladd 1990, 37–42). Major figures emerged in the debate about health, cities and planning, namely the mayors Johannes Miquel and Franz Adickes and the town planner Josef Stübben. I will return to these personalities in the next chapter.

Poverty enclaves in central city areas were at the same time a target of market, gentrification and security forces, as in the typical example of Paris in the days of Emperor Napoleon III and of the Prefect Georges-Eugène Haussmann, whose expropriation and demolition work was made possible by an 1852 imperial decree on the city's street network and then continued until 1870. A building regulation followed for the periphery of Paris.[7] Paris was transformed into a city of modern avenues, improved sewerage and water networks, markets and open parks. Residents of the central area were of course forcibly expelled. The Paris precedent had a remarkable impact in Germany (Ladd 1990, 46) (Fig. 6.1).

Developments in other European cities also made a great impression and became models to be imitated. One can mention the 1859 plan of Ildefonso Cerdà for Barcelona's expansion (Massey 2018, 137) and the 1857 plan for Vienna which included the Ringstrasse, around the old city (Schröteler-von Brandt, 2018). The radical or modest character of urban renewal projects varied from city to city and

[7] See: Jacquot et al. (2019), Orillard (2018), Bawatneh (2017), Kalflèche (2018), Massey (2018), Schröteler-von Brandt (2018), Ragon (1986), Harouel (1993), Ladd (1990).

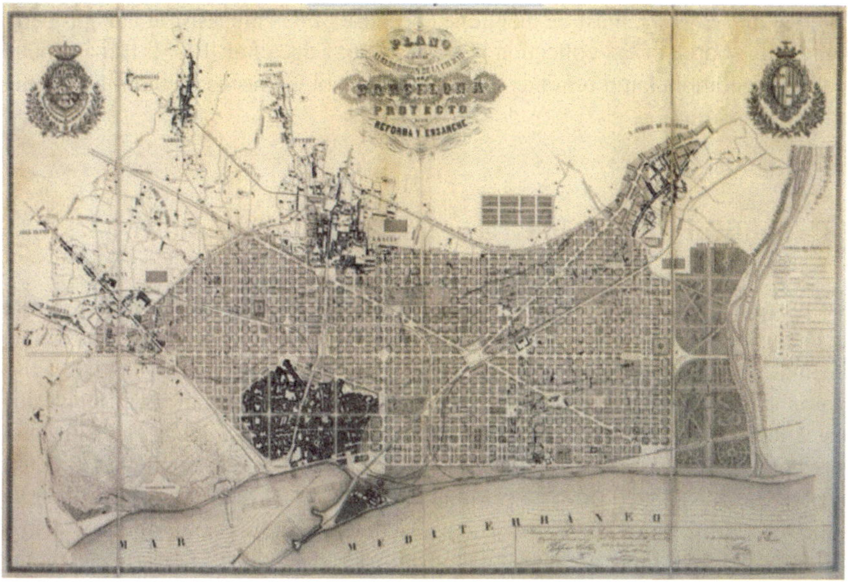

Fig. 6.1 Cerdà's plan for the expansion of Barcelona. *Source* Alberini (2021)

depended on the availability of public land and the acceptance of conservation priorities. Careful advance planning was however becoming essential. A much-discussed case was that of Berlin, where a new building regulation came to force in 1853. German cities were under different planning regimes because of the country's federal system, since 1815, which ended in 1871 when the German Reich was founded. Nevertheless, individual *länder* gradually transferred to municipalities powers of street-layout planning and land acquisition (Albers 1980, 145–146; Ladd 1990, 80; Pahl-Weber and Henckel 2008; Schröteler-von Brandt 2018). In the 1850s, a Prussian law had enabled the production of town plans for urban expansion, as with the plan of James Hobrecht for Berlin. A major step was the Prussian building alignment law of 1875 (*Preußisches Fluchtliniengesetz*) which authorized municipalities to produce binding statutory town plans. Innovative plans were designed in Cologne and Dresden, which even included an early zoning. A new town was planned next to Cologne by Josef Stübben (Ladd 1990).

New settlements were created in those years by progressive industrialists. I mention only the industrial townships of Jean Dollfuss in Mulhouse, the *Familistère* of Jean-Baptiste André Godin in Guise, Saltaire of Titus Salt in Yorkshire, and Port Sunlight of William Hesketh Lever in Merseyside. Other designs were also produced of a utopian inspiration (Choay 1965; Cherry 1996; Ragon 1986; Fernández-Armesto 2002; Kafkoula 2007). In the same period, town planning becomes the object of scientific writings. The first town planning manual was that of Reinhard Baumeister (Ladd 1990; Albers 1980), who emphasized the cooperation of architects and engineers. Although, it is not a strictly planning publication, the 1880 book by Henry George

Progress and Poverty, must be mentioned because of its emphasis on land property as the only source of tax collection (Cherry 1996; Edge et al. 1973). It may sound as an exaggeration, but it reminds me of the views of Gemistos Plethon in fifteenth century Byzantium.

6.4 Political Change in Greece and New Constitutions

The pure Bavarian regime came to an end in 1843, when a military coup led to the dismissal of Bavarians from government posts and one more national assembly (Beaton 2020, 133–134; Vournas 1974, 277, 278, 286; Petropoulos 1997, 570–578). Beaton's sarcastic comment is that the "nanny" may have left, but the real parents, the Protecting Powers, were still there. The relations between the English and King Otto were very strained. At the same time the local primates were disappointed, because they were still excluded from power (Gallant 2017, 114). Vournas stresses the role in the revolt of Colonel Kallergis, who had served in the Russian Foreign Ministry. Kallergis, before turning up to lead the rebellion, had attended a performance of the opera *Lucrezia Borgia* by Gaetano Donizetti (Petropoulos 1997, 577)! The national assembly approved a new Constitution, which however had to be ratified by a royal decree and stipulated that no one can be deprived of his property except for reasons of public need. Although supposedly following French and Belgian constitutional models, the constitution did not copy their provisions about the sovereignty of the people.[8] The rights of the king were maintained (Alivizatos 2011, 89). The constitution was his own creation in spite of some concessions for the purpose of rescuing his throne. Kaltsonis (2017, 56) adds that a balance was found between the interests of foreign powers and those of the ascending bourgeois class. Even this timid step was practically cancelled in real terms. The king was able to appoint governments as he pleased, amid a climate of electoral rigging and violence. The results of the 1844 parliamentary elections were the most disputed in the history of the country. Partisanship, corruption and opportunism were the rule (Vakalopoulos 2005, 200–201), which of course made any discussion of fair town planning a real joke.

The diplomatic relations of England and the king were almost hostile, because of Otto's irredentist views, probably caused by his desire to remain popular. When the house of Don Pacifico, an English citizen, was set on fire in 1850, possibly because he was Jewish, these relations got worse. The port of Piraeus was blockaded in 1854 by an English flotilla commanded by an officer named Parker—hence described as the "Parkerika" affair—and the British Foreign Secretary spoke of the "long arm of the Empire" and of "gunboat diplomacy" (Beaton 2020, 134–136; Vakalopoulos 2005, 206). A new popular revolt in 1862 led to the final deposal of King Otto (Alivizatos 2011, 109). The new 1864 Constitution, after one more national assembly, was signed by the new Danish King George I and stipulated that no royal decision is valid unless signed first by a minister of the lawfully elected government. It also created

[8] Internet source: filologika.gr.

a supreme advisory Council of State and repeated the need to distribute national lands. A plebiscite, the first in Greek history, had taken place before the national assembly, and approved as new king Prince Alfred, son of Queen Victoria. Russia and France objected and England proposed a Danish prince (Kaltsonis 2017, 63–64; Alivizatos 2011, 110–111). The new constitution survived for a very long period and was probably one of the most democratic constitutions of the day. It introduced the secret ballot for all males (Beaton 2020, 144). In 1875 the parliamentary vote of confidence in new governments was made compulsory (Kalyvas 2015, 51; Kaltsonis 2017, 72; Alivizatos 2011, 128–129; Kostis 2018, 356). England was obviously pleased with the final eviction of Otto. A clear proof is the concession to Greece of the Ionian Islands (McGrew 1985). In a sense, England was now linking its presence in the Eastern Mediterranean with Greece, and no longer with the Ottoman Empire.

6.5 Internal Affairs and National Lands

In parallel with constitutional changes, a number of legislative initiatives had a fundamental impact on municipal administration and local—regional and urban—economies. The 1847 law on municipal taxes allowed municipalities which were the seat of provincial authorities to impose special duties—known as "gate duties" or tariffs—on products imported by land or sea from other municipalities. The products included drinks, foodstuffs, grains, flour and "other" goods, including manufacturing products. In practice, the law introduced a division of the national territory into unequal segments and separate markets, to the benefit of those local authorities, mainly port cities, which were in a position to impose these duties (Sakellaropoulos 1991, 316–343). A later prime minister, in fact Charilaos Tricoupis, spoke of an "unheard of municipal tax". The collection of the duties was subleased to private tax collectors who had every reason to press for the perpetuation of this system, as was of course the case of local producers who avoided external competition. The gate duties institution survived for many decades and had a serious impact on income distribution, local inequalities and the increased ability of certain municipalities to fund their plan implementation.

Transport infrastructure was an expedient for a more efficient movement of goods, better access to markets, and greater equality between towns. A key project was the opening of the Corinth Canal, the beginning of which was an 1869 law. The actual work started in 1882 and continued until 1893 (Kafkoula et al. 1990, 84). The benefit for the ports of Piraeus, Patras and Corinth was enormous, at the expense of Ermoupolis on the island of Syros. One more project, of which the first steps are made in the same period, was the bridge of Evripos linking mainland Greece with the large island of Euboea. A first bridge was constructed in 1858, only to be replaced in 1890 (Demetoulis 2008). In the industrial development field, it is worth mentioning the 1864 purchase of mining rights by the foreign Roux-Serpieri firm, after a hustle regarding the reopening of the ancient silver mines of Lavrion near Athens (Veremis and Koliopoulos 2013, 130). In this case too the benefit for the economy of the Athens metropolitan area was substantial. Problems of a different nature had not ceased to exist like the cholera epidemic of 1850 (Papastefanaki 2005) or the activity of bandits, e.g., in the case of the murder of English travellers at Dilesi in 1870 (Veremis and Koliopoulos 2013, 65, 102–103; Sakellaropoulos 1991, 239). Urbanization and urban growth were still modest. As reported in the Government Gazette, the total 1856 population of Athens and Piraeus did not exceed 37,000 inhabitants.

After the annexation of the Ionian Islands, that of the region of Thessaly in 1881 was of great importance (Beaton 2020, 153–154), not only because of its rich agricultural plains, but also because of its important urban centres, like the port city of Volos. Rich merchants and bankers of the Greek Diaspora had been quick to buy land even before the annexation was finalized (Veremis and Koliopoulos 2013, 108). The radical movement which appeared among peasants could not have a serious impact at least during the nineteenth century. In the country as a whole, according to the radical critique of Vournas (1974, 437), political instability and patronage continued. Political parties lacked a clear ideological orientation and consisted of followers of individual leaders. Between 1864 and 1882 the country was ruled by 33 different haphazard government coalitions (Beaton 2020, 145), while old-style local primates were triumphant (Kostis 2018, 344). Similar bitter comments are made by Clogg (2021, 59–61), who remarks that political party cadres had simply replaced the Ottoman agas. Politics were dominated by clientelism (Kalyvas 2015, 52; Campbell and Sherrard 1968, 85, 99; Campbell 1964; Tsoucalas 1977). The power of the ruling class depended on large real estate property and on speculative investments (Vakalopoulos 2005, 232–233). The Greeks of the Diaspora did not differ from typical entrepreneurs. Although patriots, they still behaved as compradors and speculators, with strictly commercial interests (Dertilis 2019, 30–32). It can be added, that several real estate investors had been involved in party politics, which increased their power in the context of the chaotic conditions of the 1860s and 1870s.

The future of state-owned national lands was still an open question. As explained by Gallant (2017, 117), during the Ottoman occupation almost three quarters of cultivated land were in the hands of Muslims, either through outright ownership, or via the fiefdom system. During the independence war, some of the escaping Turks managed to sell their land, but in most cases simply abandoned it. The Greek revolutionary governments nationalized this land and offered it as guarantee to borrow money through international loans. At the end of the war, there existed two categories of Greek owners, i.e., those who had legally acquired their land and those who had usurped ownership rights or had received them from military commanders as a reward for their services. In Ottoman days, Sakellaropoulos (1991, 61, 69) clarifies, the Sultan granted tax and management rights over fief land to military officers, mainly the *sipahi* cavalry men. In actual fact, land was still being cultivated by the old owners, or their offspring, who had lost their ownership rights after the Ottoman conquest but enjoyed the *in rem* right of *tasarruf*, while the Ottoman state retained "bare" ownership usufruct rights. *Tasarruf* rights could be sold or inherited by relatives; they did not amount to freehold possession, but were still wider than mere usufruct rights. In the years of Ottoman decline, the Sultan granted ownership rights to wealthy Christian families, who ended up controlling almost an equal amount of land to that of Muslims (Moskof 1972, 60–65). These families were to play later a leading role in Greek politics.

After independence, *tasarruf* holders demanded full recognition of their ownership rights, but the Greek state pursued full possession, by "right of war", a claim which local Greek courts refused to recognize. Ultimately, through an 1836 law on private forests, the state achieved recognition only in the case of forests, with the exception of forest land of which the owners possessed legally valid titles. Nevertheless, in the post-war chaotic conditions the State still remained the main land owner and subcontracted its land to landless cultivators. The conditions in Attica, extensively presented by Tounta (1998), had a particular character. Even during the Ottoman rule, relative freedom was granted to the administration of local notables to facilitate tax collection. But the main characteristic was the free sale of Turkish land in Attica and Euboea, permitted by the international protocols which sealed Greek independence. The vast majority of land estates was sold to a few Greek and foreign capital holders, whose names were listed in an 1831 issue of the General Gazette of Greece.

Greek governments were faced with dilemmas carefully explained by Dertilis (2019, 158–159). The choices were to sell national land to primates and merchants or distribute them to peasants or a combination of these alternatives. The state remained undecided until 1871, on one hand because national lands were pawned as a guarantee for borrowing in international markets, and, on the other, because of the absence of a cadastral survey. However, the first excuse was often violated and the main reason was the political games of the governments. According to Veremis and Koliopoulos (2013, 83), successive governments tacitly allowed the continued use of *tasarruf* rights or tolerated land-grabbing. The whole process was bogued down and inconsistencies remained (Kostis 2018, 302–304). In various official statements, the terms "security of private property", "land interests" and "protection of citizens" are endlessly reiterated (Veremis et al. 2018. 358–359). The problems accumulate, along with the pressures of the foreign Allied Powers, who, in 1857, appoint an International Financial Commission of Inquiry) to investigate financial management and tax collection. Land distribution was entangled in wider issues of economic policy (McGrew 1985, 172). The commission continued its work until 1859, and made proposals on the repayment of a loan which Greece had contracted in 1833 as a penalty for its policies during the Crimean War. After a succession of laws in the 1850 and 1860 decades on real estate properties and the erection of buildings on national land, we reach the critical and much-awaited law of 1871 on the distribution of national lands, the possibility of which had been included in the 1864 Constitution. At last, a decision had to be made, but the timing was also linked with pressures from domestic interests.

The Greek economy was highly dependent on the production of raisins and currants, and the associated export trade to West Europe, mainly England. In retrospect, this activity may sound trivial, but in those days was very important and it was thriving partly because French raisin production was in sharp decline, as a result of a grapevine disease. In Greece, currant export traders were anxious to encourage small-scale agricultural producers to engage in vineyard cultivation and the production of raisins. If national lands were to be distributed to such producers, export trade would grow exponentially. Production was concentrated in the Peloponnese, mainly in the Corinth area, but also Messenia. It is not an accident that the decision to legislate land distribution in 1871, was taken by a government of which both the prime minister and the economics minister were deputies elected in Messenia. The law had been preceded by another act which provided incentives for new methods of producing dried currants (Sakellaropoulos 1991, 223). Land distribution was first attempted in 1864 but finally achieved seven years later, when the 1871 law distributed land to about 50.000 agricultural families and legalized dubious property titles (Veremis and Koliopoulos 2013, 129), without touching the large estates of Attica (Vakalopoulos 2005, 232). Fully aware of the enormous difference of situations, I cannot forget that at about the same time land was distributed to serfs in Tzarist Russia (Roberts 1996, 363). McGrew (1985, 207–217) offers a detailed picture of the Greek land distribution and concludes that the 1871 law finally achieved the legalization of the agricultural land, encouraged new forms of cultivation, increased the volume of commercial transactions, and improved the income of even the most remote villages. The increase of public revenues was not spectacular, especially when compared with the size of the public debt. The irony is that when French currant production recovered, the effect on Greece was such, that it probably contributed greatly to the bankruptcy of the 1890s (Fig. 6.2).

Why is currant production and the distribution of land so important for urban development and town planning? The fragmentation of land property into small parcels had a serious impact on the geography of national space. It produced the precondition of a scattered urbanization pattern, in which town planning policies proved totally ineffective. The point is made by Mantouvalou and Kalantzopoulou (2005, 87), who argue that the use of property ownership as a tool of social policy led to the multiplication of small land parcels, which gradually were transformed into building plots and gave birth to the phenomenon of out-of-plan construction, sometimes legal, very frequently illegal. Town planning proved incapable of catching up with these developments. Timid town planning efforts continued in existing towns and cities, as in Athens and Piraeus.

Fig. 6.2 Exportation of raisins from the port of Patras. *Source* G. Koutsoukou, *Come to learn history* (www.bostanistas.gr)

6.6 Town Planning for Urban Expansion, New Settlements and Building Regulations

Planning decrees for Athens and Piraeus in the period examined in this chapter had a limited corrective character or sanctioned extensions of the built-up area. As Demetrakopoulos (1934, 1937, 423) counted, from the days of the plan of Athens by von Klenze to the mid-1930s, there have been 950 plan amendments, roughly 10 per year, and 150 plan extensions. I examined a sample of amendment and extension decrees of 1850, 1851, 1863, 1865 and 1869 which confirm this ad hoc character of improvements, which apparently tended to satisfy private interests. A good example is the addition of new building blocks around Omonoia Square, which in the original Kleanthis and Schaubert plan was destined to be the site of the royal palace. It illustrates a pragmatist return to reality! Of great benefit for the quality of life of the residents was the creation, by an 1856 decree, of a zone surrounding the city of Athens, in which building was prohibited. An 1863 decree repeated this arrangement for the city of Piraeus. These zones were a predecessor of "green belts", but, in the long run, both decrees remained a dead letter. Town plan extensions facilitated the issuing of building permissions, which after 1852 were the responsibility of the so-called Administrative Police of Athens and Piraeus.

There were of course some optimists, who realized the need of a new plan for the entire city of Athens. In 1860 the mayor of Athens commissioned a study to a committee presided by an army officer of the engineering corps, of which the work proved vain (Demetrakopoulos, 1934–1937, 420). According to Biris (2005, 108–110), the committee was unduly optimistic and proved "clumsy". Apparently, the lack of reliable topographical maps was a handicap, although, in 1857, the government had invited a French mission of surveyors for a geodesic survey (Kafkoula et al. 1990, 44). It was 20 years later, in 1878, that the topographer Johann August Kaupert published a map of the Athens basin, in collaboration with an archaeologist (Kaupert, n.d.) (Fig. 6.3).

While Athens continued to expand in a random fashion without a cohesive plan, another historic development is initiated in the transport sector. In 1855, a law is approved for the construction of an Athens-Piraeus railway line, which bears the signature of the Prime Minister Alexandros Mavrokordatos, a major political figure since the days of the independence war (Papagiannakis 1990, 47–48).

Athens changes rapidly and industrial development begins to appear in timid steps. A famous example in the city's history is the beer brewery of Johann Fix of 1893, which took over from a small beer pub of 1860 in the centre of Athens (Agriantoni 2021) (Fig. 6.4).

Planning legislation is not limited to Athens. Successive decrees are published on the regulation of building activity, imposition of penalties, and demolition of buildings violating the approved street alignment or being in danger of collapse. The implementation of plans is made easier by the possibility of demolition (Monioudi-Gavala, 2017, 26–27). One of the decrees prescribed a mode or pressurizing land

Fig. 6.3 Map of Athens by Kaupert (dated 1878). *Source* Kaupert (n.d.). *Maps of Attica*

Fig. 6.4 The Athens–Piraeus railway in 1870. *Source* Stasinopoulos (1973)

owners to merge their properties into larger building plots and sell them to potential rich buyers.

Illegal construction on land occupied by trespassers remained a major problem, as it remains to some extent even today. An 1855 law determined that those who had built before 1853 houses, workshops, hotels, and windmills on national land and those legally owning such buildings had to pay to the state a sum equal to the value of these land "parcels", unless these parcels had been granted to them through the family dotation laws. However, those who had erected buildings on national land or to replace nationally-owned derelict buildings on the basis of prior legislation were excluded from the application of the new law. An elaborate procedure was introduced for the assessment of the value of land parcels. An interesting point is that, for the first time, out-of-plan "parcels" (*gipeda*) were differentiated from "plots" (*oikopeda*) in areas covered by an approved plan. The entire process proved impossible to implement and it was the 1871 law on the distribution of national lands which finally solved the problem of land ownership.

In existing towns, for which a statutory plan had been approved, the problem still existed of street layout, as well as street and infrastructure construction and maintenance. An 1856 law and an implementation decree of 1858 imposed on owners of abutting properties the duty to maintain pavements used by pedestrians and to contribute a share of the cost of sewer construction. Another decree prescribed pavement widths, construction methods and the use of materials. It is worth remembering that the construction of pavements was the first piece in the mosaic of quasi-planning legislation in sixteenth century France and the object of a decree on stone-paving of the first Bavarian period in Greece. The care for the functionality of towns, or for what the British would call amenity and convenience (Ashworth 1968, 184), reappears in Greece in an 1859 law on the use of national land for food markets in all towns and townships.

On a wider scale, an 1868 law on road construction classified roads into four categories—national, provincial, municipal, rural—a terminology still being used today. Municipal roads were those connecting the villages of the same municipality, as well as those within the limits of a town or settlement. We are approaching a stage when infrastructures become the top priority of spatial policy, as was the case in other European countries. This raises the issues of the distribution of construction costs and of the compensation of affected real estate owners. Several decrees and laws of the 1850s and 1860s tried to tackle this problem, e.g., in the case of the port city of Patras, and re-introduced the instruments of surface rights, property exchange, and compensation. Town plans were now being approved by short decrees of only a couple of lines, without explanations, as in Patras, Lamia, Nafplion and Argos. If proved unimplementable, they were hastily readjusted. A law of 1867 on the implementation of plans tried to put some order, but when reading the law, one is left with many questions regarding its reasoning and expectations. The law simply confirms the existing impasse. All-in-all, during the period 1844–1881, plans were approved for 71 towns, in addition to those for Athens, Nafplion, Argos and Ermoupolis (Demetrakopoulos, 1934–1937).

As earlier, in this same period the enactment of laws on the settlement of refugees and displaced populations continues unabated. Three laws were voted for the settlement of refugees from the island of Psara in Euboea. In 1875, several decades after the end of the independence war, a law extended the deadlines of all previous decrees. New laws and decrees were approved for three new settlements due to receive refugees from Crete. It was further proof of the vagueness which characterized all previous arrangements, even though an all-embracing law of 1866 “on settlements” had supposedly settled all disputes.

Reference has to be made to new settlement cases which were mostly related to war refugees. The new town near Corinth, after the 1856 earthquake, was a different special case. An 1858 decree approved the construction of a new town and a later law of the same year fixed the exact location, compensations, land distribution, building specifications, municipal lands, and port facilities. Yet, many years later, in 1871 and 1873, new acts were voted to accommodate those who had failed to meet the application deadline. A second case was of a coastal suburb and port of the town of Kalamata, in Messenia, which was the object of an 1860 decree. The old town was in a sense extended, but a purely agricultural zone remained between it and its coastal suburb. A third example is that of a decree of 1861 on the erection of a new town in Lokris of Central Greece, which was the predecessor of an industrial and mining community many years later. Finally, a unique case is that of the *Spaniolika* workers’ settlement in Lavrion, near Athens, of which the first plan was approved in 1867 (Kafkoula et al. 1990, 84–86; Kafkoula 2007, 191 και 226). It was the first company town in Greece, which owes its name to the quarters of Spanish workers hired by the mining enterprise of Illarion Roux for the re-smelting of ancient metallurgical slag remnants. The residential quarters were located next to the processing plant of the firm of Roux-Serpieri-Freissynet C.E.,[9] where the workers were accommodated in 1865. An 1873 decree approved the exploitation of mines in various places, including that of Lavrion. A further case of workers’ quarters which will be mentioned in the next chapter was that of those employed in the drainage project of Lake Copais. The project was approved by an 1867 law.

6.7 Conclusion

I can describe the period examined in this chapter as transitional. After the feverish activity of the Capodistrian years and of the first years of the Bavarian administration, and before the premiership of Charilaos Tricoupis, one is left with the impression that a state of fatigue prevails. The problems that are dealt with are mostly leftovers from the past, something which is particularly evident in town planning affairs. The constitutional revisions opened new horizons, of which, however, the results were slow to appear. Greece remains static, with the exception of the annexation of the Ionian Islands. Governments are immersed in the conditions inherited from the past and continue to be subjected to the pressures of dependence on the Protecting Powers and to the effects of their inability to solve old problems, like the distribution of national lands.

Town planning is reduced to patching up initiatives, mainly plan amendments and ad hoc urban extensions. Most of the efforts for new town creation tend to be limited to old obligations. There is everywhere a sense of rotten apples and stagnation, in spite

---

[9] Historical Archive of Lavreotiki Municipality. istoriko.lavreotiki.gr.

of the first steps of some infrastructure projects. Legislative initiatives continue, but they are static and repeat or amend older laws, e.g., on road construction or building regulation. With the exception of the law on national land distribution, achieved at last by a breathless state after years of procrastination, there is no law which was a real turning point. Even that law was a response to specific commercial interests of the period. Its long-term side-effects were to lead to the perpetuation of out-of-plan building activity. Town planning continues to depend on the historic 1835 decree on the sanitary construction of towns, and on narrowly targeted regulatory acts for Athens. A sterile mentality dominates, which is devoid of imagination and audacity, when, at the same time, other countries engage in innovations, albeit sometimes of an extreme character. When one attempts a comparison with the achievements of earlier or later periods, he is bound to observe a shortage of ideas and creative actions.

References[10]

Agriantoni C (2021) German technological superiority in Greece, 1900–1940: practices and conceptions. In: Kyrtsis AA, Pechlivanos M (eds) Compendium of Greek–German encounters. URL: https://comdeg.eu/el/compendium/essay/104279/ [GR]

Alberini C (2021) The relationships between port and city in the Mediterranean basin: the cases of Barcelona and Genoa. PORTUSplus. J RETE N. ISSN: 2039-6422

Albers G (1980) Town planning in Germany: change and continuity under conditions of political turbulence. In: Cherry GE (ed) Shaping an urban world. Mansell, London, pp 145–160

Alivizatos N (2011) The constitution and its enemies in modern Greek history, 1800–2010. Polis, Athens [GR]

Ashworth W (1968) The genesis of modern British town planning, 1st edition 1954. Routledge & Kegan Paul, London

Bawatneh S (2017) Planning of Paris city (1853–1870)/case study. Research Gate. https://doi.org/10.13140/RG.2.2.13828.78720

Beaton R (2020) Greece: Biography of a modern nation. Penguin Random House, London

Benevolo L (1967) The origins of modern town planning. Routledge & Kegan Paul, London (Italian original 1963)

Bergin J (2015) A history of France. Palgrave, London

Bernstein W (2009) A splendid exchange: how trade shaped the world. Atlantic Books, London

Biris K (2005) Athens: From the nineteenth to the twentieth century, Revised edition. Melissa, Athens [GR]

Bourillon L (2000) La loi du 13 avril 1850 ou lorsque la Seconde République invente le logement insalubre. Revue D'histoire Du XIXe Siècle 2000(20/21):117–134

Brogan H (2001) The penguin history of the USA. Penguin, London

Brunn SD, Williams JF (1983) Cities of the world. Harper & Row, New York

Campbell JK (1964) Honour, family and patronage. Clarendon Press, Oxford

Campbell JK, Sherrard P (1968) Modern Greece. Ernest Benn, London

Cherry G (1996) Town planning in Britain since 1900: the rise and fall of the planning ideal. Blackwell, Oxford

Choay F (1965) L'urbanisme: Utopies et réalités—Une anthologie. Seuil, Paris

Choay F (1969) The modern city: planning in the nineteenth century. Studio Vista, London

Clogg R (2021) A concise history of Greece. Cambridge University Press, Cambridge

[10] References in Greek are marked with the initials GR.

Demetoulis K (2008) Pothitos Kamaras and his contribution to the digging of the Evripos strait. In: Technical Chamber of Greece—Euboea Regional Annex. 45 years of the drawer-bridge of Evripos. Album. Chalkis, pp 19–37 [GR]
Demetrakopoulos A (1934–1937) Town plans. Techniki Epetiris tis Ellados, 1934–1937, vol B (II), 359–449 [GR]
Dertilis G (2019) History of modern and contemporary Greece, 1750–2015. Revised edition. Crete University Press, Herakleion [GR]
Drian J-C (n.d.) 1850–1995—Les étapes de la politique du logement en France. Temporalités. https://journals.openedition.org/rh19/212
Edge G et al (1973) The city as an economic system. The Open University Press, Bletchley (UK)
Fernández-Armesto F (2002) Civilizations. Touchstone—Simon & Schuster, New York
Flitouris LA (2015) Colonial empires. Asini, Athens [GR]
Gallant TW (2017a) Modern Greece: from independence to the present, 1st edition 2001. Bloomsbury Academic, London
George P (1964) Précis de géographie urbaine. Presses Universitaires de France, Paris
Hall P (1998) Cities in civilization. Weidenfeld & Nicolson, London
Hardoy JE (1975a) Preface. In: Hardoy JE (ed) Urbanization in Latin America. Anchor Books, Garden City (NY), pp vii–xix
Hardoy JE (1975b) Two thousand years of Latin American urbanization. In: Hardoy JE (ed) Urbanization in Latin America. Anchor Books, Garden City (NY), pp 3–55
Harouel J-L (1993) Histoire de l'urbanisme. Presses Universitaires de France—Collection Que Sais-je? Paris
Heap D (1969) An outline of planning law, 5th edn. Sweet & Maxwell, London
Hugueney J (1950) Un centenaire oublié: La première loi française d'urbanisme. Vie Urbaine, 13 avril 1850, 241–249
Jacquot H, Priet F, Soazic M (2019) Droit de l'urbanisme. Dalloz, Paris
Jenkins S (2012) A short history of England. Profile Books, London
Kafkoula K (2007) The adventure of garden cities. University Studio Press, Thessaloniki [GR]
Kafkoula K, Papamichos N, Hastaoglou V (1990) Town plans in Greece of the nineteenth century. Thessaloniki [GR]
Kalflèche G (2018) Droit de l'urbanisme. Thémis Droit/Presses Universitaires de France, Paris
Kaltsonis D (2017) Constitutional history of Greece, 1821–2001. ΚΨΜ, Athens [GR]
Kalyvas SN (2015) Modern Greece. Oxford University Press, New York
Kaupert JA (n.d.) Maps of Attica. Trochalia and Glyfada Municipality, Athens [GR]
Kostis K (2018) "The spoiled children of history": the making of the modern Greek state, eighteenth–twentieth centuries. Patakis, Athens [GR]
Ladd B (1990) Urban planning and civic order in Germany: 1860–1914. Harvard University Press, Cambridge (Mass)
Legge JD (1980) Indonesia. Prentice Hall of Australia, Sydney
Lichtheim G (1971) Imperialism. Allen Lane—The Penguin Press, London
Mak G (2008) In Europe: travels in the twentieth century. Greek translation. Metaichmio, Athens (Dutch original 2004) [GR]
Mantouvalou M, Kalantzopoulou M (2005) Town planning and sociopolitical stakes in interwar Greece. In: Eleftherios Venizelos and the Greek city: town planning policies and socio-political re-arrangements. Conference proceedings, Chania, 24–27 Oct 2002. National Foundation of Research and Studies "Eleftherios Venizelos", Technical Chamber of Greece and National Technical University of Athens, Athens pp 85–95 [GR]
Massey D (2018) Key planning histories of the developing Western tradition from the mid-nineteenth century to the early twentieth century. In: Hein C (ed) (2018) The Routledge handbook of planning history. Routledge, London, pp 134–146
McGrew WW (1985) Land and revolution in modern Greece 1800–1881. The Kent State University Press, Kent (Ohio)

Monclús J, Díez Medina C (2018) Urbanisme, urbanismo, urbanistica: Latin European urbanism. In: Hein C (ed) (2018) The Routledge handbook of planning history. Routledge, London, pp 147–160
Money D (1976) South America. University Tutorial Press, Slough
Monioudi-Gavala D (2017) Planning and real estate property in Athens, 1833–1922. Paraskinio, Athens [GR]
Monteil V (1972) Indonésie. Éditions du Seuil, Paris
Moskof K (1972) National and social conscience in Greece,1830–1909: the ideology of compradorist space. Thessaloniki [GR]
Orillard C (2018) Urbanisme and the Francophone sphere. In: Hein C (ed) (2018) The Routledge handbook of planning history. Routledge, London, pp 161–172
Pahl-Weber E, Henckel D (eds) (2008) The planning system and planning terms in Germany: a glossary. Academy for Spatial Research and Planning (ARL), Hanover
Papagiannakis L (1990) The Greek railways (1882–1910), 1st edition 1982. National Bank Cultural Foundation, Athens [GR]
Papastefanaki L (2005) Public health, tuberculosis and professional pathology in Greek cities of the early twentieth century: The contradictory process of urban modernization. In: Eleftherios Venizelos and the Greek city: Town planning policies and socio-political re-arrangements (2005). Conference proceedings, Chania, 24–27 October 2002. Athens: National Foundation of Research and Studies "Eleftherios Venizelos", Technical Chamber of Greece and National Technical University of Athens, Athens, pp 155–170 [GR]
Pendle G (1967) A history of Latin America. Penguin, Harmondsworth
Perrot JC (1992) Genèse d'une ville moderne: Caen au VVIIIe siècle (première publication: 1974). In: Roncayolo M, Paquot Th (eds) (1992) Villes et civilisation urbaine: XVIIIe–XXe siècle. Larousse, Paris, pp 34–46
Petropoulos IA (1997) Politics and statecraft in the kingdom of Greece, 1833–1843. Greek translation. National Bank Cultural Foundation. Athens (original in English 1968; see JA Petropulos) [GR]
Phalip L (2019) Histoire de l'urbanisme. Bréal, Paris
Pinol JL (2000) Le monde des villes au XiXe siècle. Greek translation. Plethron, Athens [GR]
Prescott WH (1979) History of the conquest of Mexico and history of the conquest of Peru. Single volume, 1st editions 1843 and 1847 respectively. The Modern Library, New York
Ragon M (1986) Histoire de l'architecture et de l'urbanisme modernes, vol 1. Idéologies et pionniers 1800–1910. 3 volumes. Casterman, Paris
Roberts JM (1996) A history of Europe. Helicon, Oxford
Rodriguez R (2005) The foundational process of cities in Spanish America: the Law of Indies as a planning tool for urbanization in early colonial towns in Venezuela. Focus 2(1):47–57
Sakellaropoulos ThD (1991). Institutional transformation and economic development: State and economy in Greece, 1830–1922. Exantas, Athens [GR]
Schröteler-von Brandt H (2018) The history of urban planning. Academy for Territorial Development in the Leibniz Association (ARL). https://nbn-resolving.org/urn:nbn:de:0156-559 97441
Stasinopoulos EK (1973) History of the Athenians: from antiquity to our age. Athens [GR]
Tombs R (2015) The English and their history. Penguin, London
Tounta F (1998) Landed property, urban expansion and deforestation in Attica: the case of the forest estate of Stamata and Dionysos. Doctoral thesis, National Technical University of Athens, National Archive of Doctoral Theses (www.didaktorika.gr) [GR]
Trevelyan GM (1952) History of England, 1st edition 1926. Longmans, Green & Co, London
Tsoucalas K (1977) The problem of political clients in nineteenth century Greece. In: Kontogiorgis GD (ed) Social and political forces in Greece. Greek Society of Political Science & Exantas, Athens, pp 73–112 [GR]
Vakalopoulos KA (2005) History of Greece/Single volume synthesis: from the genesis of new Hellenism (1204) to modern Greece. Ant. Stamoulis, Thessaloniki [GR]

Veremis ThM, Koliopoulos GS (2013). Modern Greece: a history from 1821. Patakis, Athens [GR]

Veremis ThM, Koliopoulos GS, Michailidis ID (2018) 1821: The creation of a nation-state. Metaichmio, Athens [GR]

Vournas T (1974) History of modern Greece: From the revolution of 1821 to the revolt of Goudi of 1909. Tolidis Bros, Athens [GR]

Wassenhoven L (2019) The ancestry of regional spatial planning: A planner's look at history. Springer, Cham (Switzerland)

Williamson E (2009) The Penguin history of Latin America. Penguin, London

Chapter 7
The End of the Nineteenth Century: The Emerging Town Planning Science and Developments in Greece

Abstract After the annexation of Thessaly, a period follows in Greece during which pure town planning issues retreat to second place, in comparison with public works, infrastructures, and the production sector. Under a new modernizing political leadership, the country attempts to change its industrial profile, an effort which is violently interrupted by a major financial bankruptcy and a military defeat just before the end of the nineteenth century. All these developments are reviewed in this chapter. The disastrous bankruptcy, the imposition, once again, of foreign financial control, and the slow pace of industrialization prevent the country to modernize and to introduce town planning innovations, at a time of major changes in the discipline and practice of planning of western European countries. This is the reason why in this chapter the author refers to influential writers like Josef Stübben and Camillo Sitte and to the dialogue that ensues in Europe. The "science" of town planning is consolidated, the planning profession grows and professional associations are being formed. In Greece, while Athens and the cities of Thessaly are rapidly changing, planning is limited to endless town extensions and ad hoc amendments of existing layout plans.

Keywords Infrastructures · Industrialization · Bankruptcy · Town planning science · Plan amendments

7.1 Introductory Note

On an international scale this is the period when town planning comes of age. From the perspective of a town planner who follows developments in Greece from the annexation of Thessaly in 1881 to the military humiliation of 1897, one could claim that the period is uninteresting. This conclusion would be based on the examples of town planning, which are limited to plan amendments and plans for the suburbs of Athens. However, in terms of the restructuring of urban and regional space, the conclusion would be misleading. The reasons are the rapid changes in the productive, economic and technological systems, the construction of public infrastructures, the signs of modernization, in spite of its interruption by a fiscal bankruptcy, and the

L. C. Wassenhoven, *The History of Town Planning in a Dependent State*,
https://doi.org/10.1007/978-3-031-96958-4_7

rising maturity of society, which is entering a new age. The entire framework within which town planning operates changes radically. A further cause of the changing town planning context, as the nineteenth century reaches its end and the next century begins, is that in industrialized countries the nature of planning as science and practice is being redefined and gains social respect. This evolution is channelled to Greece in many ways and affects the thinking of politicians and practicing planners.

7.2 The Rising Face of Town Planning on an International Scale

Countries going through a phase of intense industrialization were bound to pay attention to its effects on cities, urbanization trends, and deteriorating housing conditions. Employment patterns were rapidly changing as rural labourers, urban employees in the service sector, and small-scale manufacturing workers sought jobs in large industrial plants. This seismic change was not limited to advanced countries like France and Germany, with Britain being in the vanguard. In early nineteenth century, in Germany, the Napoleonic wars led to a complete urban restructuring and even to the abolition of most free cities. It was however in Imperial Germany in the late century that cities grew rapidly (Fulbrook 2004, 74, 79, 97,137; Roberts 1996, 318). As Peter Hall (1975, 22–23), reminds us with reference to the phenomena of the new industrial town and of massive rural–urban migration. industrialization did not affect only the urban network of West European countries. In Central Europe too, as Gorzelak (1996, 38) points out, nineteenth century industrialization was a force, along with long-run historical trends, which produced the settlement structure and the spatial organization of the twentieth century. Countries which nineteenth century industrialization left relatively untouched, in comparison to the rest of Europe, e.g. in the Balkans, had a different task. They were trying, according to Yerolympos (1996, pp.14–15), to modernize, in fact "europeanize" the cities which they had inherited from the period of Ottoman rule. She gives us a survey of developments in Romania, Serbia, Bulgaria and Greece, countries mostly concerned with extensive urban renewal. In Greece, apart from the planning of Athens, plans were also produced for ancient and historic towns, like Sparta or Eretria (see also Karadimou-Yerolympos 2004).

The influence of European developed countries on Greece in the period under examination in this chapter was of course a continuation of the conditions of earlier years, with Britain being now the dominant force. Foreign control determined the country' political orientation and its ability to contract much-needed loans. This control reached a peak when Greece declared bankruptcy in 1893 and made the foolish decision to declare war on Turkey in 1897.

Compared to these dramatic developments, town planning is certainly of lesser importance, but here too external influence remained decisive. The attention paid to town planning in advanced European countries was gradually imitated in Greece, not only because of the involvement of foreign planners in Greek affairs, but also because Greek architects had studied or worked abroad, mainly in Germany and France. As pointed out by Karadimou-Yerolympos (1995, 41), in those days, progress in planning was made primarily in England, Germany, France and the United States of America. She devotes a special chapter to these countries and to the written work of Baumeister, Josef Stübben and Sitte, with which I will deal later.

In spite of the presence in Greece, though after World War I, of planners like the Frenchman Ernest Hébrard and the Englishman Thomas Mawson, it was probably nineteenth century developments in Germany which had a wider influence. English planning historians recognize it emphatically. Cherry (1996, 32–33) remarks that, although a new planning movement in all four of the above countries was evident, it was in Germany that an elaborate planning system was created thanks to the mobilization of municipal authorities, and then recalls the admiration for this system of the British observer T.C. Horsfall. Houghton-Evans (1978, 54) too accepts Germany as a model, refers to Horsfall, and underlines the German invention of zoning.

This is a period when the metropolitan centres of London, Paris, Berlin and Vienna are in the midst of a veritable population explosion (Mak 2008, 43, 52), which demands new regulatory initiatives. Pioneering experiments are also visible in lesser cities. Such efforts are supported by legislative innovations in quick succession. They concern local government, housing, hygiene, aesthetics, layout of streets and building blocks, and building regulations. They also introduce instruments like zoning and land adjustment. A leading figure of German innovations was Franz Adickes, mayor of Frankfurt from 1891 to 1912 (Ladd 1990, 10, 17), repeatedly mentioned by Greek planning historians (Kafkoula 2007, 144–145; Karadimou-Yerolympos 1995, 108–110; Yerolympos 1996, 29). The 1891 Adickes plan for Frankfurt preserved the character of the Altstadt, the old town, but included a full zoning arrangement. The plan was approved, and later the curved street-pattern and the maintenance of the urban fabric were reproduced in many German cities (Ladd 1990, 131). Ladd describes in detail the living conditions and housing patterns in Berlin, Cologne, Düsseldorf, and Frankfurt. It is generally recognized that Frankfurt opened the way for housing reform in Germany and that Adickes was a real activist (*ibid.*, 151–170).

This was a time of intensive inter-state interaction, even across the Atlantic. German reformers had been inspired by the work of Henry George in America (*ibid.*, 177). In France, the 1894 law Jules Siegfried, forced the administration to deal with social housing. Napoleon III was aware that older attempts to provide housing for workers had failed because of high land costs and housing rents. It was for this reason that he proceeded to land acquisition in advance and subsidies to workers (Ragon 1986, 115). The improvement of working-class conditions was now squarely on the town planning agenda. McAuslan (1980, 3) is quick to point out that the roots of modern town planning can be found in these appalling conditions. For their improvement, new powers were needed for the regulation and control of real estate properties, land use, and housing quality. These powers, when introduced in England by law, were no doubt resented by property owners, who appealed to the courts of law. Common law judgements of the late nineteenth century continued to apply until after World War II, due to the continued importance of housing conditions. The recognition of German town planning achievements was extended to the housing field. German reformists had for years placed emphasis on the freedom of the housing market, but also on the creation of building cooperatives which would ensure the access to decent housing of working-class households (Logan 1976, 378). In contrast, for the high housing costs and the excessive residential densities, liberals blamed land taxes, building regulations and the limitations imposed on the free movement of capital and labour. Nevertheless, progressive reforms were gradually accepted.

It should not escape our attention that a close bond existed between housing and health policies, in order to check the spread of infectious diseases. Hamburg suffered in 1892 by a cholera epidemic (Ladd 1990, 20). High densities and poor sanitation favoured contamination. Land use control and zoning were instruments of sanitary policy. Zoning was in effect a German invention of the late nineteenth century. Its first experimental use represented an effort to improve housing conditions by regulating density. Although it was sometimes accompanied by attempts to achieve a functional organization of urban areas, it was not yet the instrument which evolved after World War I. Yet, zones exclusively destined for housing had been introduced by law already in the 1860s and 1870s (*ibid*., 187). The public benefit was used as a criterion. Although its value as a yardstick was often disputed, it now appeared increasingly in the arguments of public authorities. It is of interest that a French law of 1883 required that careful research must be carried out before a project is considered as contributing to the public benefit (Phalip 2019, 92). The contribution of zoning to public health objectives was gradually established in the logic of town planning, although disagreements continued to persist, as Ladd (1990, 189–195) shows in his analysis. The requirements of better lighting and ventilation implied lower buildings, larger courtyards, and, if possible, free-standing houses. In his 1876 textbook, Reinhard Baumeister recommended density variations and functional diversification in cities. He and Adickes supported the integration of zoning in building regulations. Baumeister's view was that zoning was a compromise between hygiene objectives and economic reality. The first zoning regulation was adopted in Frankfurt in 1891. Appeals were lodged in the courts of law, but were rejected. However, in real terms the regulation was not successful as expected because of property speculation and the expectation of windfall profits.

Zoning regulations were introduced later in other cities as well, e.g., in Cologne where a plan by Josef Stübben included four zones of varying density. Resistance to zoning gradually subsided and zoning ordinances became more complex. Alternative approaches to check speculation in the property market were also put to the test, helped by improvements in cadastral records (Hertel and Wicke 2004). In the meantime, zoning had been exported to the United States (Logan 1976, 377; Sutcliffe 1981, 210). Its potential had impressed American planners, including Benjamin C. Marsh, author of the textbook *An Introduction to City Planning*. An English planner, namely Houghton-Evans (1980, 106), confirms that the most widely accepted planning idea of the nineteenth century was land-use zoning. Zoning proved valuable for the protection of architectural heritage and cityscapes. Such was the case with the French law of 1887 for the protection of monuments and historic buildings (Jacquot et al. 2019, 31–32; Gohier 1965). According to Logan (1976, 378), German and Austrian architect-planners, like Baumeister, Sitte, Josef Stübben, Gurlitt, Goecke, Henrici, Fischer, and Hegemann, specially admired in the United States, paid attention to the urban fabric and aesthetics.

Urban plans were now becoming the object of several German planning competitions (Ladd 1990, 132–137). Planning made increased use not only of zoning, but also of the instrument of land adjustment to make possible the implementation of building regulations. This was the purpose of planning ordinances in Baden and Hamburg. Many municipal authorities were now imposing zones of different types of buildings (Schröteler-von Brandt 2018). Of a comparatively conservative nature was the 1894 London Building Act for the public control of street-layout, building alignments, open spaces and building heights (Ashworth 1968, 181). A factor of great significance in Germany was the versatility and freedom of action of local authorities. Influential legislation was of course enacted in other countries as well, like the municipal administration act of 1884 in France which introduced the obligation to produce layout plans (Jacquot et al. 2019, 30; Gohier 1965). In England, after the 1835 law on Municipal Corporations, a law on public health was voted in 1875, in accordance to which 700 city health agencies were created. The most important innovation came with an 1888 law for the establishment of County Boroughs, independent of the old County Councils (Cherry 1996, 1–4).

I referred earlier to the tool of land adjustment, another most important German innovation. It is worth a comment because we encountered an early version, although not named as such, in Greece at the time of Governor Capodistria, and was also used in the reconstruction of the city of Thessaloniki after the 1917 fire. Doebele (ed., 1982) defines it as a method which enables property owners of an area to participate in a process of urban development and share its gains. This definition fails to convey the essence of restructuring and redistributing land properties. The terms used to describe this instrument are reparcelling, land readjustment or simply land adjustment. The method is frequently applied in exceptional circumstances, after natural disasters (Larkham 2018, 431). Land adjustment was used in Germany following the enactment of a Prussian law, known as *Lex Adickes*, and was finally implemented only in Frankfurt. As explained by Ladd (1990, 200–201), land adjustment (*Umlegung*) was used by municipalities willing to expropriate land in areas with very small land building lots, retain a land percentage for opening streets, and redistribute the rest to property owners in the form of buildable plots. Land adjustment was the solution in areas of property fragmentation, which did not permit development, unless small plots were united into co-owned properties. Against the opposition of many proprietors, Mayor Adickes managed to convince the Prussian parliament to allow land adjustment in Frankfurt, although its use had already been tested in Hamburg, Baden and Saxony. It was extended much later to other Prussian cities, including Cologne, although its application proved slow and cumbersome.

7.3 Planning Redefined and Garden City Movement

Planning innovations are propagated not only by copying successful plans or reproducing model solutions, transferred from other countries or cities. They are also inspired by academic studies, addressed to new generations of planners. I have already mentioned the work of Reinhard Baumeister. In the last years of the nineteenth century two very influential books appeared, first by Camillo Sitte in 1889, then that by Josef Stübben in 1890, and a few years later the book by Ebanezer Howard. Even the term *Städtebau* was introduced in the books by Sitte and Stübben. Until then terms like "town expansion" were being used (Albers 1980, 146). Let us not forget, that in Greece the term *chorotaxia* (space-ordering) was a translation of the German equivalent *Raumordnung* (Doxiadis 1942, 1). German terminology is explained by Celina Kress (2018, 173–175), who reminds us that a definition of town planning was given in 1904 by Goecke and Sitte in the journal *Der Städtebau*: "*Städtebau* is the unification of all technical and creative arts into a great enclosed entity … *Städtebau* is a science. *Städtebau* is a kind of art with specific aims of research, and specific big tasks for practical execution… in which numerous technicians, artists, national economists, hygenicists, social politicians, civil servants and lawyers are working together… with all inhabitants of our cities participating" (*ibid.*, 173).

In Britain, as pointed out by Cherry (1996, 17), the term "town planning" appears for the first time around 1906. The practice of conscious town building, according to which the constitutive elements of the town form are defined in accordance with deliberate design principles, is millennia-old and has been presented in a host of studies. Even without precise definitions in planning law, a new activity in urban affairs now emerges in a short time-span, based on key ideas. Around this nucleus of concepts, Cherry concludes, a new corpus of knowledge was formed, which secured its presence in political and institutional structures.

As with town planning, the term *urbanisme* was first used in France at the end of the nineteenth century. It meant, Orillard (2018, 161) explains, and continues to mean, more than just town design. It englobes a vast field of sciences, technologies and arts related to the structure and development of urban space, in order to enhance human wellbeing and social cohesion, while preserving the environment. Most francophone historians identify *urbanisme* with an "urban science", which appears around 1900 and is akin to town planning of the anglophone countries and to *Städtebau* of the German tradition. A number of French professional planners were now making a distinction between *urbanisme* and old practices, e.g., those of the time of Haussmann, which was later labelled *art urbain*. Planning historians now speak of the much more complex nature of *urbanisme*, which was born out of a process of international exchange of ideas. The planning profession is now the offspring of this interaction. It was in this environment that the French society of town planners was created in 1911 (Gohier 1965).

In parallel with the debate about the scientific identity of planners, legislative reforms continue in France with laws on public hygiene and sanitary regulations, which municipalities and prefectures had to impose on building contractors, or on the demolition of dangerous buildings (Jacquot et al. 2019, 30–31). Another law, at the beginning of the twentieth century dealt with natural landscape and cityscape protection. An 1887 act had already ordered the protection of historic and listed buildings. Kress (2018) argues that in Germany the integration in actual planning practice of the new meaning of *Städtebau* was compatible with the tradition of townscape (*Stadtlandschaft*) protection and the organismic modelling of cities.[1]

At the turn of the century, new directions are also apparent at the level of building regulations. Schröteler-von Brandt (2018) singles out the 1900 building regulation of Saxony and its zoning provisions. The zones it prescribed, roughly similar to those listed in the contemporary federal building regulation of Germany *(Baugesetzbuch)*, were copied, with variations, by various cities. Soon they became the foundation of public civic design at the local level in cities like Frankfurt, Mannheim and Munich. In actual fact, the Saxony building regulation was the first civic design law in Germany.

As the nineteenth century is about to end, the international dialogue on town planning is enriched by the 1898 book by Ebenezer Howard on Garden Cities (Ashworth 1968, 138–142; Choay 1965, 277–289; Corden 1977, 39–43; Hall and Ward 1998). It is rare for a book in the planning field to have such a lasting impact, of which evidence is provided by a publication in Greek on garden cities more than a century later (Kafkoula 2007). A year after Howard's book the Garden City Association was formed in England. According to F.G. Osborn's foreword to the 1965 edition, the book had a visionary brilliance. Yet, it was criticized by conservatives as mere phantasy and by progressives as full of over-simplistic panaceas. Even some Fabian Society members were of the opinion that the book had no practical value. Howard himself thought that there is no effective medicine for the cities, unless attractive solutions are made available to the people (Howard 1965, 45–48). He offered his famous illustration of the "city-country-people" magnets and he appealed for the movement of people from congested cities to mother earth. The city-magnet offers opportunities and enrichment, at a high cost of hard work, inhuman slums and social exclusion. The country-magnet offers natural beauty, real wealth, clean air and high life quality. The city symbolizes social symbiosis, while nature cares for the human being. City and country must forge a complementary existence to foster hope, life and culture. Garden cities have to combine these qualities.

Only five years later, in 1903, the first garden city was founded, i.e., Letchworth Garden City, on a plan by Barry Parker and Raymond Unwin (Miller 1981, 79; Ashworth 1968, 142–143; Abercrombie, 1967, 98–99). The Hamstead Garden Suburb in London followed in 1906, on a plan by Raymond Unwin (Miller 1981, 83). The next garden city had to wait until 1920. Welwyn Garden City was built at the initiative of Howard himself, with Louis de Soissons as architect-planner.

This was no longer the time for the founding of capital cities like those mentioned in earlier chapters, but, as colonial empires were still surviving, I can mention a famous example, although its foundation belongs to the twentieth century. The accession to the British throne of King George V was marked by the removal in 1911 of

[1] At this point she refers to the work of the Greek architect Papageorgiou-Venetas.

the capital of the British Raj in the Indian subcontinent from Calcutta to Delhi, a move which was “one of the best-kept secrets of modern times” (Spear 1978, 179). The transfer was due to the activity of nationalist movements in Calcutta. New Delhi, which was finally inaugurated in 1931, was largely designed by the architects Edwin Lutyens and Herbert Baker. Delhi was of course an ancient city and New Delhi is but a small part of the Delhi conurbation (King 2007, 234–235; Brunn and Williams 1983, 353–354).

Collins and Grasemann Collins (1965, 20) recognize the influence of Howard and his ideas on German planning and observe that the picturesque and informal character of garden cities were particularly welcomed by the Germans. Schröteler-von Brandt (2018) links the garden city reception with municipal decentralization. However, Howard’s ideas survived only in some urban periphery housing developments in the vicinity of Essen, Berlin and Wittenberg. An imitation of the garden city model can be found in some later suburbs of Athens, like Psychiko or Ekali, which were mostly developed by private cooperatives and became high-class residential areas, a far cry from Howard’s urban–rural prototype. Their building plots were gradually purchased by members of the high-income class of Athens or by foreign diplomatic missions. Psychiko is located at a fairly small distance from the centre of Athens, an advantage which increases its attractiveness. A traffic plan commissioned by the municipality imposed a large number of one-way streets to discourage through-traffic. Individual building plots are still in demand for the construction of luxury villas or apartment blocks. Many embassies and ambassadorial residences, with the exception of those located in the city centre, are now to be found in Psychiko, as they are in a separate district of the Australian capital Canberra of which the construction began in 1913 on a design much influenced by the garden city idea (Figs. 7.1 and 7.2). With a minor exception, land uses for commercial or professional premises are not allowed in the main body of Psychiko. Recently the Psychiko municipality was merged with that of the neighbouring Filothei suburb, yet another high quality housing area. The Psychiko suburb (still referred to as Old Psychiko) is located near New Psyshiko

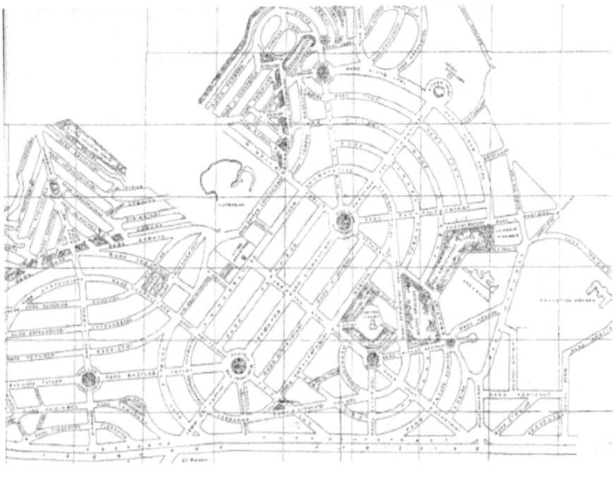

Fig. 7.1 Street layout plan of Psychiko suburb in Athens. *Source* Municipality of Psychiko

Fig. 7.2 1932 aerial photograph of Psychiko. *Source* Municipality of Psychiko

(now part of the same municipality), which is a commercial and services area to which Old Psychiko residents turn for shopping and a broad range of services.

The circular patterns of the Psychiko street layout remind me of a phrase in a television serial about Paris: "Do you realize that Paris is laid out in circles, like they deliberately designed it to confuse us".

7.4 Planning Textbooks

A most interesting aspect of planning is the interaction between practice and academic writing. Three books written in German during this period are excellent examples of feedback loops between these facets of planning work. They were produced by Reinhard Baumeister, in 1876, already mentioned in the previous chapter, Camillo Sitte and Josef Stübben, in 1889 and 1890 respectively. It is better to start with Stübben, because Sitte's positions were to a large extent responses to Stübben's earlier professional work as a consultant. The second reason is that Stübben revised later his views on the protection of the traditional townscape and came closer to Sitte's ideas. His book was reprinted in a re-edited version in 1907 and 1924. Stübben's and Sitte's work has been repeatedly quoted and analysed by several authors. In the view of Starostenko (2019, 438–440), the book by Stübben (*Der Städtebau*), is considered the first major work on the theory of planning and an invaluable manual for many generations of planners. Because it contains a wealth of practical examples, it has been described as a cooking guide, full of planning recipes. The earlier book of Reinhard Baumeister, was a forerunner of both Sitte and Stübben, who emphasized different aspects of town planning. Josef Stübben focused more on the layout of the street network, while Sitte cared more for the aesthetic dimension of town planning, which he considered an art. A caricature of their difference, is that Stübben preferred rectangular grids and straight streets, while Sitte wanted curved streets and ever-changing vistas. In Starostenko's opinion, between the first and the second edition of Stübben's book there is a shift of interest from building blocks and housing to zones of work and production. There is however another fundamental change in relation to the protection of traditional urban environments.

The differences reflect ideas which had emerged between the editions of 1890 and 1907. This is not only a matter of Stübben's change of attitude. It marks a wider change of concerns shared by a wider community of citizens, who were conscious of traditional values and reacted against the Haussmann "bulldozer" mentality, which was to return several decades later after World War II. The themes of the chapters of Stübben's first edition are highly indicative: typology and construction of houses, transport, street classification, place of public buildings, elements of design, building blocks, typology of squares, water surfaces, and plan extensions. However, in the second edition, a chapter was added on old city quarters. It is clear that now the rehabilitation of historic towns was approached with a qualitatively different mind. Stübben now included sections on town planning history, town morphology, functional variations, and compactness v. openness of town development. A faith conversion, not to speak of repentance, is hidden in this change. This conversion is brilliantly discussed by Ladd (1990, 122–131). It was, he argues, the period in which the value of the urban fabric is rediscovered. In about a decade, the views of Sitte had become dominant in Germany, e.g., in Frankfurt, Ulm or Cologne. Sitte's emphasis on the visual and spatial relationship of buildings, monuments, streets and squares encouraged the supporters of townscape and building conservation. In 1893, only three years after the first publication of his book, Stübben remarked that no one would now agree with the demolition of a medieval gate in Cologne, which had been sacrificed 11 years earlier for the sake of vehicle traffic. What caused this remark was his appeal to protect the remnants of a Roman gate which was about to be knocked down for the same reason. Josef Stübben was the best example of the conversion of opinion-makers of his day. According to Ladd, he was not only the most famous planning theoretician and professional of the 1890s, but also something like a weather vane, because he tended to follow the most imaginative and innovative theoreticians and popularize, and then adopt, their ideas. In the 1870s he was charmed by Haussmann's Paris, although he was aware of the critique against the destruction of the city's fabric. His design for Neustadt in Cologne had followed the Haussmann style of radial and diagonal streets and monumental vistas. Yet, in the discussions of the Cologne municipal council for the Roman gate he had become a faithful partisan of conservation. The Haussmann followers of 1879 must have been astounded. His conversion takes place during the period of the consolidation of the planning profession in Germany, although the established taboos were hard to eradicate.

The effect of Sitte's thinking on Josef Stübben is difficult to measure, but what is certain is that Sitte was a second pole of ideas. Collins and Grasemann Collins (1965, 26 και 30) stress that the German vanguard role in town planning was based on the work of Stübben, Baumeister and Sitte, but also on the theories of aesthetics of Hermann Maertens and Charles Buls. In their opinion, Sitte's critique was directed mainly against the technicians and experts. He did not condemn a particular profession, but rather the mentality which was dominant among architects, engineers and municipal officials. He was not the only person to voice criticism. There were others who felt that at the root of the problem were the engineers, given that they played a major role in German town planning, in contrast to the United States where the presence of landscape architects was more frequent. In Germany, planning departments were controlled by engineers. Sitte was particularly critical of surveyors, because they were only interested in recording the use of flat surfaces. He was giving priority to the vertical component of town planning, so as to highlight the third dimension of buildings. The sort of town plan (*Bebauungsplan*) he advocated became the basis of more strict and detailed building regulations, particularly as far as height zoning was concerned, an issue to which German town planners paid increasing attention in the 1890s. Various commentators had described the bureaucratic priorities of officials in planning departments, who failed to place due attention to soil morphology and urban forms. Sitte despised the geometer-surveyor who cared only for geometrical precision and ruler-dominated town planning; he also rejected the stereotypical solution of solid building blocks and triangular or asteroid public squares.

These designs served another principle which angered Sitte, i.e., the priority given to traffic planning, either in new towns, or in the restructuring of old ones. Horse-drawn carriages were determining the design of streets and squares. Traffic engineers in the modern sense did not exist in those days, but town planners, like Baumeister and Stübben, were known for the importance they attached to traffic. Several commentators of the time wrote that curvilinear and meandering streets were picturesque, but not suitable for the movement of carriages. The views of Baumeister and Stübben on vehicular traffic explain naturally their admiration for the French school of town planning which favoured the model of Paris. According to the followers of the Sitte school, another priority of German planners which made them servants of the engineers, was the emphasis on hygiene and sanitation. With the spread of infectious diseases, as I showed earlier, a kind of town planning variant grew, which concentrated on sewage disposal and underground water control. Sitte was angered because he thought that infrastructure planning tended to undermine the free and organismic character of town planning above the earth's surface (Collins and Grasemann Collins 1965, 16–19).

Sitte, in the view of the same authors, had been influenced by Aristotle and was frequently referring to the nature of the city, in an aesthetic and psychological sense. He was interested in the perception of the city through the eyes of the observer, not on the basis of an *ex ante* plan. For him the city was an architectural creation conceived in three dimensions. In a sense, Sitte was giving town planning back to the architects. Town planning was the arrangement of urban space in attractive and useful forms, not just the lining up of building blocks along traffic arteries which Josef Stübben advocated. Sitte's admirers insisted that he had turned the art of designing cities into a real art of space (*Raumkunst*). For him the constituent elements of an approach aiming at city as a holistic work of art were the concepts of "architecture and nature" and the variations of alternating "solids and voids" (*ibid.*, 45–53). The change which Sitte was calling for was, in Ladd's opinion, the incorporation in the practice of town expansion planning of aesthetic criteria and, also, the rejection of the prevailing engineering perception of spatial quality. He gave prominence to the relationship of historic buildings and monuments with their environment and he opposed the tactic of their decongestion and of the demolition of all surrounding structures, which prevented the open vistas to the monuments. Sitte himself argued that if the artistic approach could be embedded in town planning and if rich artistic forces could be attracted to this task, possibly through planning competitions, then more could be achieved, even though the vision of the ideal city would remain beyond our grasp (Sitte 1992, 167–168).

The issue itself of the restoration of historic and traditional buildings, Ladd adds, does not appear in Sitte's book, but it certainly affected decisions regarding the location of important buildings. It reinforced the rising perception of the architectural framework and could not be divorced from it, in full contrast to the views of the first planners, who, when confronted with problems of sanitation and traffic, refused to accept any attempt to determine aesthetics through building regulations. In this respect, Baumeister's theses were a reflection of the liberal climate of the 1870s. Baumeister did not consider aesthetic objectives as an issue of public needs requiring state regulatory guidance, and trusted rather the market forces to produce attractive buildings. As to the late—repented—Stübben, he had come to agree that the statutory regulation of building heights, floors, protrusions, and forms, which had been popular in the first half of the century, was wrong in comparison with the new ideas of building beauty (Ladd 1990, 111).

The books by Baumeister and Josef Stübben were never translated in Greek. Sitte's book was translated by Serraos almost a century after its initial publication (Sitte 1992). In his foreword to the Greek edition, Aravantinos reminds us that Sitte had visited Greece in 1889. It is impossible to determine the degree to which this fascinating international debate had an impact on Greece, or any other country for that matter. However, its reverberations certainly reached the ears of a number of planners, particularly those who had studied or worked in Germany, in the early years of the twentieth century, like Costas Kitsikis, Antonis Kriesis and Anargyros Demetrakopoulos, and then played a significant role as university professors or as consultants in important urban planning activities or public agencies. The same remark can be made for other planners who were educated, e.g., in France, like Marinos Delladetsimas. Equally, as I mentioned earlier, the work of Ebenezer Howard had a spreading influence which found its expression in quasi-garden city experiments by way of its acceptance in Germany or France. The truth is that in the years right after the publication of the books by Stübben, Sitte and Howard, the political and economic conditions in Greece were not propitious, because of the public attention to national infrastructure works, an economic crisis, military adventures, and political upheavals.

7.5 Domestic Conditions

In the period 1881–1897 Greece is a different country in many respects. Territorially, it is now substantially larger, but the main differences are due to the change of its development orientation, which affects its social structure and the settlement network (Beaton 2020, 154).

The personality of Prime Minister Charilaos Tricoupis, his reforms and the fiscal debacle also had a major influence. Vakalopoulos deals in detail with Tricoupis's internal reconstruction of the state and emphasizes the priorities of his government. These priorities included a strengthened parliamentary system, clear evidence of his British experience, the principle of majority rule, the performance of the civil service, administration reform, and improved transport facilities. After the Capodistria period, Tricoupis was probably the most important nineteenth century political personality of the country (Vakalopoulos 2005, 288–291). Although the political chaos of previous decades had been left behind, the conflict between Tricoupis's modernizing party and the national party of Deliggiannis, with its irredentist proclamations, perpetuated a state of tension. The government, dependent as it was on high spending and borrowing, walked on a tightrope (Beaton 2020, 164). The public debt was restructured in 1878 and gave the government further access to international markets, but the need to invest in military armaments and in infrastructures remained pressing and Tricoupis was struggling to increase investment in both. The situation deteriorated further during the so-called currants' crisis of 1892 and bankruptcy was unavoidable. In 1893, Tricoupis admitted openly that the country was bankrupt. A much more important cause was the constant military preparations, which culminated with the folly of the 1897 war with Turkey, under the administration of Prime Minister Deliggiannis. Extreme patriotic sentiments and chauvinism had taken over and the economy was derailed (Dertilis 2019, 473; 2016, 53, 59), at a time of a prolonged international economic crisis of the period 1873–1896 (Tsoulfidis 2015, 218). During the 1897 war, the Ottoman forces reached the suburbs of Athens and it was the intervention of European powers—not just the three Protecting Powers of Britain, France and Russia, but also Germany, Austria and Italy—which saved Greece. A compensation had of course to be paid to Turkey and the Greek economy was once again placed under the supervision of an international financial control committee (Dertilis 2016, 63; Beaton 2020, 167–170). The irony is that in 1896 Athens had hosted the first modern Olympic Games (Roidis 2004). A most significant development was the creation in 1898 of the autonomous state of Crete, under the supervision and control of European powers. Ultimately, Crete joined the Greek state in 1913.

The 1881 annexation of the agricultural region of Thessaly had serious side-effects and political repercussions, which were related to the ownership of land and the profile of the owners. Within a year, 75% of the land had been acquired by foreign-based Greeks. The new owners were merchants and bankers. Thus, the large rural estates are in a sense assimilated to the urban economy (Moskof 1972, 136–138). Public land-ownership is radically reduced and conditions for the peasant-cultivator were totally altered, because of the character of the new owner, who functioned as a long-distance investor (Sakellaropoulos 1991, 93–94, 102–103). Following this development, large properties become the rule in Thessaly, as they had been in Attica and Euboea, while in other regions midget properties are dominant (Tounta 1998). The developments in Thessaly imply new relations between urban investment capital and agricultural production and between cities and countryside. At the same time a transformation occurs in urban economic activities, city development profiles, capital accumulation, urbanization processes, and core-periphery relationships (Demathas and Tsougiopoulos 1985, 508–510). The new capital investment forms can be detected in legislation, as in an 1893 decree on commercial and industrial patents and trademarks, which were undoubtedly the result partly of private enterprise pressures (Sakellaropoulos 1991, 229), partly of government policies to encourage large-scale, viable production units. That was the year of construction of the Fix brewery plant, which became a landmark of the Athens agglomeration. Its owner, of Bavarian descent, was of course anxious to secure a monopoly of beer production (Agriantoni 2021; Anastassopoulos 1947, vol. II, 799). The government's intention was clearly to leave behind the model of the single-product agricultural economy—like the raisin production and export—and enter the industrial age (Tsoulfidis 2015, 200–201). In the view of Tsoulfidis, the economic policy of the Tricoupis administration revolved around three axes: Infrastructures (ports, railways, land drainage), enhanced state credibility in international markets, and modernization of the taxation system to increase state revenues. Industrial activity was indeed escaping from the small family model of handicraft or agricultural processing units and turning towards larger plants of flour and olive oil mills, tanneries, and cotton textile mills (Anastassopoulos 1947; GGE, n.d., vol. XXIII, 324; Tsoulfidis 2015, 209–210). New laws, e.g., in 1882, regulated mineral extraction, and cigarette production, while the first tobacco manufacturing plants appear in 1883 and 1887. Tariffs on imported goods were imposed to help local manufacturing firms. New banking facilities are also introduced, with the use of promissory notes. Sanitation measures are imposed on production firms and a new Hygiene Society is created, with a wide membership not only of physicians, veterinarians, chemists, pharmacists, and naturalists, but also of architects, engineers and journalists. Here too, the influence of German and French precedents is evident.

7.6 Public Works

French and other foreign engineers were invited by Tricoupis in 1982 to assist in the major effort of public works, mainly in the transport sector (Kafkoula et al. 1990, 46). A share company of public and private works was also constituted by decree, in order to attract private funds. During Tricoupis' second term of office, an 1887 decree fixed the remuneration or foreign engineers for their work in road and bridge building. Among other public works, it is the construction of a road and, in particular, rail network which has an exceptional importance. 3000 kms of public

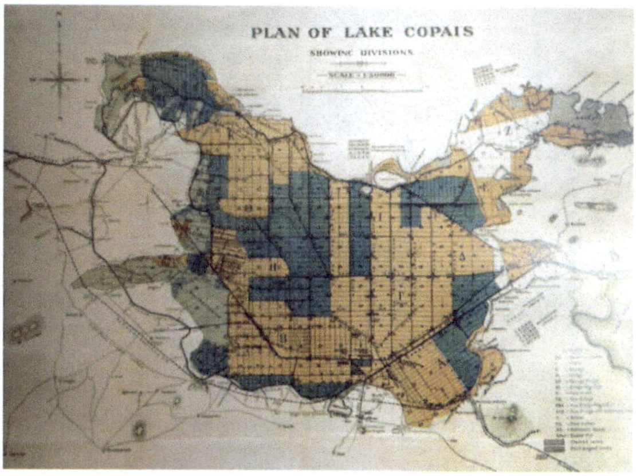

Fig. 7.3 Copais Lake: agricultural land parcelling. *Source* Maistrou et al. (2003)

roads and 1500 kms of rail lines were constructed. Considering the small size of the country, these numbers are really impressive. Tricoupis insisted that Greece is not allowed to wait any longer and estimated that the project could be finished in 10 years at most at low cost, but he was proved wrong (Tsoulfidis 2015, 202). In a special study of the history of Greek railways, Papagiannakis recounts the course of this effort (Papagiannakis 1990, 73–134), but confirms the failure of Tricoupis' prediction (*ibid.*, 10). In the case of the railways, neither the plan, nor the political will were missing, he adds. However, the modernization of transport and production infrastructures was not sufficient to achieve the take-off of the economy and overcome the inertias of Greek society. Nevertheless, in the period of Tricoupis started the construction of three rail lines, one from Piraeus to the Peloponnese, a second from Athens to Thessaly, and a third one within Attica, from Athens to the mineral town of Lavrion. Their construction took 25 years. In spite of the delay, they proved a basic component of the development of the country.[2] A law and a decree of 1882 approved the construction contract of the Peloponnese rail line and the constitution of a share company charged with its management. Another law approved the contract for the Lavrion-Athens line, which was extended to the north of the capital, up to suburb of Kifissia. In the 1880s, legislation was also enacted for road construction, hydraulic works, and the staffing of public administration with engineers and surveyors.

Government initiatives were not limited to rail and road building projects. Of great practical and symbolic significance were the drainage of Copais Lake in Boeotia, port infrastructure improvements, the deepening and widening of the shipping strait of Evripos, and, above all, the digging of the Corinth Canal (Tsoulfidis 2015, 202) (Fig. 7.3).

Legislation was passed in 1882 on the drainage of lakes, specifically the Copais Lake. Work started in Copais by a French company, but in the aftermath of floods the drainage rights were transferred to Lake Copais Co. Ltd. Drainage work was completed in 1892. The company had erected an administration centre and a small

[2] Gale A (2023) The early history of Greece's railways. Greekreporter.com.

Fig. 7.4 The small "company town" in Aliartos. *Source* Ministry of Culture

housing estate, a kind of mini-company town, adjacent to the township of Aliartos (Maistrou et al. 2003) (Fig. 7.4).

Of much greater historic significance was the Corinth Canal, which produced a new maritime shipping condition and affected favourably the ports of Piraeus and Patras, with a negative impact on that of Ermoupolis (Loukos 1985). However, Tsoulfidis (2015, 202) remarks not only that the project was over-ambitious and expensive, but also that it disappointed its initiators and those who financed it. Because of the construction delays, a decree of 1887 extended the deadline and the official inauguration took place in 1893 (Beaton 2020. 163; Kostis 2018, 462). Of importance for the communication between Attica and Boeotia with the island of Euboea and the town of Chalkis was the Evripos bridge. The dismantling of the old bridge was decided in 1894, when tenders were invited for the construction of a new revolving bridge which became operational in 1896 (Demetoulis 2008, 2009, 24–27; see also Technical Chamber of Greece—Euboea Regional Annex 2008).

7.7 Town Planning in Greece

Town planning initiatives were either due to continuing trends of urbanization and expansion of existing settlements, or, mainly, to the annexation of Thessaly, with its important towns, and to the policies of the Tricoupis government. Important factors were the processes of road and railway construction, commercialization of agriculture, and transition of the rural economy from self-consumption to market-oriented production (Kotea 2011). Towns, with the exception of Athens, could now be classified into import–export port centres—Patras, Piraeus, Ermoupolis –, minor seaboard towns, and inland centres, which now had become nodes of agricultural production trading, linking Athens with their rural hinterland.

The addition of Thessaly to the national territory had a considerable effect on the urban network, which was not noticeably altered by the railways (Kafkoula et al. 1990, 45, 83–86). The opening of the Corinth Canal, apart from its effects on shipping routes, was also the cause of two minor company towns, which housed construction workers. Important health spas, endowed with hot water springs, are also appearing in this period, in Attica, Euboea, and West Peloponnese (Garavelas 2025).

The production of town plans was prolific, but on the whole lacked originality and was limited to plan amendments and town extensions. Naturally, new plans had to be designed for the towns of Thessaly. Demetrakopoulos (1934–1937, 377–378) counted 108 town plans between 1864 and 1897. This is an impressive number, but according to him, most of these plans were deficient and full of inaccuracies. In the list he has compiled, one detects several cases of towns in Thessaly, in or near the Athens complex, as well as health spas. In Athens itself amendments continue as usual, but an interesting example is that of decrees of 1886 and 1887 approving plans to accommodate the Athens-Lavrion rail line and the new railway stations in the centre of the capital. This is a case where town planning is synchronized with transport policy. Of interest are also plans for fashionable suburbs, as Kifissia in the north of Athens or Phaleron on the coast of the Saronicos gulf.

On the whole, planning legislation of the Tricoupis period, beyond the approval of layout plans, was very limited. There was however a law on grants of public land in approved settlements, a residual of older obligations, and a law on the sanitary conditions of houses, in fact on the construction of cesspools. History repeats itself! It is worth making the point that industrial development would lead eventually to planning legislation in the interwar years, after World War I. The growth of industries and manufacturing precincts would generate problems to which town planning would have to respond. But, in the closing years of the nineteenth century this process was still at its infancy. Genuine manufacturing urban zones do not yet exist and industry is still fully dependent on agriculture and the processing of rural production (Kalogris and Tsokopoulos 1985, 431–436). Industry is concentrated in the Athens-Piraeus conurbation and some middle-sized towns, to which, after 1881, Volos is added and quickly develops into the second industrial city of the country, after Piraeus. Its maritime trade with European ports played a significant role (Papagiannopoulou 1985, 488–489). Ermoupolis followed a reverse path of decline (Agriantoni 1985, 603–604).

For Volos and the other towns of Thessaly, a priority of the Tricoupis government was to produce plans, provide the necessary infrastructures, "hellenicize" the character of these old Ottoman centres, and create an environment which would favour modern economic activity. Hence, town plans were prepared and approved from 1882 to 1904 by municipal engineers (Hastaoglou 2006, 180–183). Hastaoglou's analysis brings to mind the precedents of other European countries where efforts were made to impose order on the urban fabric, rationalize the street network, and modernize the traditional urban structure, which of course was not an Ottoman inheritance. In Greece, there was also the additional desire to escape from an urban organization created during the long years of foreign occupation. However, if we leave aside this particular pursuit, we rediscover all the objectives of street layout improvement, infrastructure provision, and house building regulation. A quick succession of laws and decrees aimed at the attainment of these objectives in all cities of Thessaly.

An event of great importance for the history of town planning in Greece took place a year after the end of the nineteenth century. I refer here to the town planning law of 1901 of the autonomous Cretan State (*Kritiki Politeia*), which I will briefly discuss in the next chapter. The events of the reconstruction of the newly liberated city of Thessaloniki, after the destructive fire of 1917, take us even further into the twentieth century, but it is worth remembering that a previous fire of 1890, when Thessaloniki was still under Turkish occupation, had also led to considerable damage and to a planning intervention (Karadimou-Yerolympos 1995, 25–31; see also Karadimou-Yerolympos 2013, 2016).

7.8 Conclusion

Two developments were given special attention in this chapter. The first was the consolidation of town planning as science and art, thanks mainly to the writings and the professional activity of major figures like Josef Stübben, Camillo Sitte and Ebenezer Howard. The second was the modernizing work in Greece of the Charilaos Tricoupis government. Apart from the pioneering contributions of the abovementioned writers, planning innovations continued in Germany, Britain and France. In Greece, progress in the town planning field was destined to be achieved more slowly, mainly as a consequence of the improvement of transport infrastructures, which slowly but steadily transformed geographical space and the role of cities, particularly Athens. It is true that change was slowed down because of the collapse of state finances and military adventures. In parallel with public works, a very hesitant industrial development takes place and an underlying social transformation can be detected, mainly in the capital city, sometimes of a disputable nature. These early signs were due to become more visible later, when society was to receive a major shock after the Asia Minor Disaster of 1922. Excessive government borrowing, fiscal instability, and inflation caused a wave of outmigration, which deprived Greece of a valuable labour force. On the other hand, when Thessaly became part of the national territory, Greece was blessed with a valuable agricultural hinterland. New dynamic urban centres were also acquired, which potentially could produce town planning innovations. The granary of Thessaly was sadly the terrain of unfortunate property rearrangements, intense urbanization, uncontrolled land subdivision, and land rental value redistribution. Public infrastructure works had the twin objective of increasing the cohesion of national space and promoting the exploitation of natural wealth sources. Railway and road construction, as well as the opening of the Corinth Canal were directed to the first objective; projects like the drainage of Lake Copais or the reuse of Lavrion mines were to help the attainment of the second. Certain cities benefit from these actions, but others are negatively affected.

On the international scene we witness a fever of town planning innovations, namely the increased emphasis on housing problems, hygienic conditions, building regulations, land use zoning, municipal powers, public benefit priorities, control of land speculation, urban infrastructures, and cadastral surveys. New towns are built, although their overall importance cannot be compared to that of existing cities. Planning competitions are on the increase and the content of planning is no longer limited to street layout designs. Planning instruments are multiplied as in the case of zoning and land readjustment. The German *Städtebau* concept becomes a model to be imitated and is praised in other developed countries. The notions of both *urbanisme* and town planning are transformed. Town planning is now the object of intense scientific and political debates. Greece is not yet the proper environment for such a debate, but the latter's effects were due to reach it in the years to come. The production of endless amendments of existing town plans continues, along with urban extensions, often related to transportation projects. Planning remains closely related to traditional practices of compromise with local vested interests. In Athens, plan extensions in suburban space are a reflection of the dynamic growth of the city and of the demand of the affluent classes for attractive second home suburbs. Yet, in other cases, plan extensions are due to the emergence of industrial centres. The dynamic character of these developments was soon to be confronted with the reality of twentieth century conflicts.

References[3]

Abercrombie P (1967) Town and country planning. In: Rigby Childs D (ed). Oxford University Press, London (originally published 1933)

Agriantoni C (1985) The transformations of the industrial structure of Ermoupolis in the nineteenth century. In: SSNH/Society for the study of new Hellenism (1985b) The Neo-Hellenic city: Ottoman heritage and Greek state. Conference proceedings, vol II. Athens, 26–28 Sept 1984, and Ermoupolis, 29–30 Sept 1984. Athens, pp 603–608 [GR]

Agriantoni C (2021) German technological superiority in Greece, 1900–1940: practices and conceptions. In: Kyrtsis AA, Pechlivanos M (eds) Compendium of Greek-German encounters. URL: https://comdeg.eu/el/compendium/essay/104279/ [GR]

Albers G (1980) Town planning in Germany: change and continuity under conditions of political turbulence. In: Cherry GE (ed) Shaping an urban world. Mansell, London, pp 145–160

Anastassopoulos GA (1947) History of Greek industry, 1840–1940, 3 volumes. Elliniki Ekdotiki Etaireia, Athens [GR]

Ashworth W (1968) The genesis of modern British town planning, 1st edition 1954. Routledge & Kegan Paul, London

Beaton R (2020) Greece: Biography of a modern nation. Penguin Random House, London

Brunn SD, Williams JF (1983) Cities of the world. Harper & Row, New York

Cherry G (1996) Town planning in Britain since 1900: the rise and fall of the planning ideal. Blackwell, Oxford

Choay F (1965) L'urbanisme: Utopies et réalités—Une anthologie. Seuil, Paris

Collins GR, Grasemann Collins C (1965) Camillo Sitte and the birth of modern city planning. Phaidon Press, London

Corden C (1977) Planned cities: New towns in Britain and America. Sage, Beverly Hills

[3] References in Greek are marked with the initials GR.

Demathas Z, Tsougiopoulos G (1985) The Neo-Hellenic city: Its development process—a theoretical framework. In: SSNH/Society for the study of new Hellenism (1985b), The Neo-Hellenic city: Ottoman heritage and Greek state. Conference proceedings, vol II. Athens, 26–28 Sept 1984, and Ermoupolis, 29–30 Sept 1984. Athens, pp 505–510 [GR]
Demetoulis K (2008) Pothitos Kamaras and his contribution to the digging of the Evripos strait. In: Technical Chamber of Greece—Euboea Regional Annex (2008) 45 years of the drawer-bridge of Evripos. Album. Chalkis, pp 19–37 [GR]
Demetoulis K (2009) Pothitos Kamaras: 1861–1935. Technical Chamber of Greece—Euboea Regional Annex [GR]
Demetrakopoulos A (1934–1937). Town plans. Techniki Epetiris tis Ellados, 1934–1937, vol B (II), 359–449 [GR]
Dertilis, G (2016) Seven wars, four civil wars, seven bankruptcies: 1821-2016. Polis, Athens [GR]
Dertilis G (2019) History of modern and contemporary Greece, 1750–2015. Revised edition. Crete University Press, Herakleion [GR]
Doebele WA (ed) (1982) Land readjustment: a different approach to financing urbanization. Lexington Books, Lexington (Mass)
Doxiadis K (1942) About spatial planning (chorotaxia). Technika Chronika, vol 241–244 [GR]
Fulbrook M (2004) A concise history of Germany. Cambridge University Press, Cambridge
Gale A (2023) The early history of Greece's railways. Greekreporter.com
Garavelas L (2025) The development of spas and their present role: the case of Eastern Central Greece. Lecture at the one-day conference "Inter-disciplinary research of Greek places with thermal springs and of their development prospects". National Technical University of Athens, 2025 [GR]
GGE (Great Greek Encyclopaedia) (n.d.) Phoenix—Pavlos Drandakis, Athens [GR]
Gohier J (1965) L'évolution de l'urbanisme en France. CRU65—Études et Essais. Centre de Recherches d'Urbanisme, Paris
Gorzelak G (1996) The regional dimension of transformation in Central Europe. Jessica Kingsley, London
Hall P (1975) Urban and regional planning. Penguin, Harmondsworth
Hall P, Ward C (1998) Sociable cities: the legacy of Ebenezer Howard. Wiley, Chichester
Hastaoglou V (2006) The towns of Thessaly, before and after 1881: the traditional town and its modernization planning. In: Thessaly, historical issues, vol I. Thessaly ETEDK, Larissa, pp 169–199 [GR]
Hertel C, Wicke H (2004). Real property law and procedure in the European Union/National Report: Germany. European University Institute (EUI) Florence/European Private Law Forum, in cooperation with Deutsches Notarinstitut (DNotI), Würzburg
Houghton-Evans W (1978) Planning cities: Legacy and portent. Lawrence & Wishart, London
Houghton-Evans W (1980) Schemata in British new town planning. In: Cherry GE (ed) Shaping an urban world. Mansell, London, pp 101–128
Howard E (1965) Garden cities of to-morrow. Edited with a preface by F.G. Osborn, 1st edition 1898. Faber & Faber, London
Jacquot H, Priet F, Soazic M (2019) Droit de l'urbanisme. Dalloz, Paris
Kafkoula K (2007) The adventure of garden cities. University Studio Press, Thessaloniki [GR]
Kafkoula K, Papamichos N, Hastaoglou V (1990) Town plans in Greece of the nineteenth century. Thessaloniki [GR]
Kalogris P, Tsokopoulos V (1985) Industry and cities in Greece of the second nineteenth century half. In: SSNH/Society for the study of New Hellenism (1985b) The Neo-Hellenic city: Ottoman heritage and Greek state. Conference proceedings, vol II. Athens, 26–28 Sept 1984, and Ermoupolis, 29–30 Sept 1984. Athens, pp 431–438 [GR]
Karadimou-Yerolympos A (1995) The reconstruction of Thessaloniki after the 1917 fire. University Studio Press, Thessaloniki [GR]
Karadimou-Yerolympos A (2004) Between East and West: Thessaloniki and northern Greece towns at the end of the nineteenth century. University Studio Press, Thessaloniki [GR]

Karadimou-Yerolympos A (2013) The emergence of modern Thessaloniki: histories, personalities, landscapes. University Studio Press, Thessaloniki [GR]
Karadimou-Yerolympos A (2016) Fortified military camp, defenseless city: disaster and transformation of Thessaloniki after the 1917 fire. University Studio Press, Thessaloniki [GR]
King AD (2007) Colonial urban development: culture, social power and environment, 1st edition 1976. Routledge, London
Kostis K (2018) "The spoiled children of history": the making of the modern Greek state, eighteenth–twentieth centuries. Patakis, Athens [GR]
Kotea M (2011) The rise and fall of mayoral power in Greece, 1833–1936. Int J Humanit Soc Sci 2011(1):13–20
Kress C (2018) The German traditions of Städtebau and Stadtlandschaft and their diffusion through global exchange. In: Hein C (ed) The Routledge handbook of planning history. Routledge, London, pp 173–191
Ladd B (1990) Urban planning and civic order in Germany: 1860–1914. Harvard University Press, Cambridge (Mass)
Larkham PJ (2018) Disasters: recovery, re-planning, reconstruction, and resilience. In: Hein C (ed) (2018) The Routledge handbook of planning history. Routledge, London, pp 428–440
Logan Th-H (1976) The Americanization of German zoning. J Am Inst Planners 42(4), 377–385
Loukos Ch (1985) A Greek city in decline: Ermoupolis in the second half of the nineteenth century. In: SSNH/Society for the study of new Hellenism (1985b) The Neo-Hellenic city: Ottoman heritage and Greek state. Conference proceedings, vol II. Athens, 26–28 Sept 1984, and Ermoupolis, 29–30 Sept 1984. Athens, pp 585–589 [GR]
Maistrou El et al (2003) Agro-industrial complex of Aliartos in Boeotia. Cultural foundation of the Piraeus Group, Athens [GR]
Mak, G (2008) In Europe: travels in the twentieth century. Greek translation. Metaichmio, Athens (Dutch original 2004) [GR]
McAuslan P (1980) The ideologies of planning law. Pergamon Press, London
Miller M (1981) Raymond Unwin: 1863–1940. In: Cherry GE (ed) Pioneers in British planning. The Architectural Press, London, pp 72–102
Moskof K (1972) National and social conscience in Greece,1830–1909: the ideology of compradorist space. Thessaloniki [GR]
Orillard C (2018) Urbanisme and the Francophone sphere. In: Hein C (ed) The Routledge handbook of planning history. Routledge, London, pp 161–172
Papagiannakis L (1990) The Greek railways (1882–1910), 1st edition 1982. National Bank Cultural Foundation, Athens [GR]
Papagiannopoulou E (1985) The port of Volos at the end of the nineteenth century: its development in relation to maritime and terrestrial transport networks. In: SSNH/Society for the study of new Hellenism (1985b), The Neo-Hellenic city: Ottoman heritage and Greek state. Conference proceedings, vol II. Athens, 26–28 Sept 1984, and Ermoupolis, 29–30 Sept 1984. Athens, pp 487–498 [GR]
Phalip L (2019) Histoire de l'urbanisme. Bréal, Paris
Ragon M (1986) Histoire de l'architecture et de l'urbanisme modernes, vol 1. Idéologies et pionniers 1800–1910. 3 volumes. Casterman, Paris
Roberts JM (1996) A history of Europe. Helicon, Oxford
Roidis E (2004) Texts on Athens of the 1896 Olympic games. In: Demetroulis D (ed) Metaichmio, Athens [GR]
Sakellaropoulos ThD (1991). Institutional transformation and economic development: State and economy in Greece, 1830–1922. Exantas, Athens [GR]
Schröteler-von Brandt H (2018) The history of urban planning. Academy for Territorial Development in the Leibniz Association (ARL). https://nbn-resolving.org/urn:nbn:de:0156-559 97441
Sitte C (1992) Der Städtebau nach seinen Künstlerischen Grundsätzen. Greek translation by K. Serraos. National Technical University of Athens, Athens (German original 1889) [GR]

Spear P (1978) A history of India, vol 2. Penguin, Harmondsworth
Starostenko Y (2019) The J. Stübben's Book Der Städtebau (town planning) and its influence on the works of Russian specialists in cities improvement of the early twentieth century. In: International conference on architecture: heritage, traditions and innovations (AHTI 2019), vol 324. Advances in social science, education and humanities research, Atlantis Press, pp 438–440
Sutcliffe A (1981) The history of urban and regional planning: an annotated bibliography. Mansell, London
Technical Chamber of Greece—Euboea Regional Annex (2008) 45 years of the drawer-bridge of Evripos. Album. Chalkis [GR]
Tounta F (1998) Landed property, urban expansion and deforestation in Attica: the case of the forest estate of Stamata and Dionysos. Doctoral thesis, National Technical University of Athens. National Archive of Doctoral Theses (www.didaktorika.gr) [GR]
Tsoulfidis L (2015) Economic history of Greece. University of Macedonia Press, Thessaloniki [GR]
Vakalopoulos KA (2005) History of Greece/Single volume synthesis: from the genesis of new Hellenism (1204) to modern Greece. Ant. Stamoulis, Thessaloniki [GR]
Yerolympos A (1996) Urban transformations in the Balkans (1820–1920). University Studio Press, Thessaloniki

Chapter 8
Early Twentieth Century

Abstract While Greece is in a stalemate, a fresh wind blows from the new Cretan State, where a major town planning legislative initiative appears in 1901. In Greece, following a military coup against the royal regime, a new liberal political movement emerges and an innovating spirit takes over under a very competent leadership. In the West, it is the time of major changes in planning ideas, expressed, e.g., in the work of Ebenezer Howard and Patrick Geddes. Planning in Britain, France and Germany is on a new path, in terms both of legislation and its orientation in original directions. These new developments are the subject of this chapter. In Greece the effect of new directions is delayed by the Balkan Wars of 1912–1913, its involvement in the Great War, the integration of the regions of Epirus, Macedonia and Thrace, the disastrous 1922 Asia Minor war against Turkey, and the displacement to Greece of 1,3 million refugees. In spite of these dire conditions, the liberal opening to new ideas, the cooperation of French and English planners, and the work of Greek planners educated in the West, leads to the Legislative Decree of 1923 on the plans of cities, towns and settlements. The tragedy of a wildfire which ravaged the centre of the newly liberated city of Thessaloniki also caused the introduction of new planning methods.

Keywords 1923 planning decree · Foreign influence · Refugee problem · Thessaloniki wildfire · Balkan wars · National schism

8.1 Introductory Note

The events of the first quarter of the twentieth century are of course beyond the ambit of this book, but they are in a sense an extension of prior developments and deserve our attention. Their origins are found in the nineteenth century and their manifestation can only be appreciated when they are seen as a delayed outcome of past trends. The role of European powers, especially Britain, continues to be a decisive factor and does not appear out of the blue. The autonomous Cretan State was born before 1900, but its effect on Greece was to be of a long duration, in political life, but also in town

L. C. Wassenhoven, *The History of Town Planning in a Dependent State*,
https://doi.org/10.1007/978-3-031-96958-4_8

planning, because of Crete's 1901 planning law. A key figure of the Cretan State, Eleftherios Venizelos, was soon to become the leader of the Greek Liberal Party and Prime Minister of Greece.

The decline of the Ottoman Empire continues and its transformation into the new state of Turkey is a development closely related to events in Greece. What happens in Greece is not a follow-up of the period of Tricoupis. The wars of the 1910 decade dramatically extend the frontiers of the country to the north. Town planning enters a new stage with the reconstruction of Thessaloniki, now part of the national territory; a stage of which the culmination is the 1923 town planning legislative decree. Developments in Germany, France and England exercise deep influence on planning, as new innovative thinking is derived from the writings of authors like Patrick Geddes and as distinguished foreign planners work in Greece, like the French Ernest Hébrard and the English Thomas Mawson. Planning legislation in France, England and Germany established new standards, which Greeks emulate. The arrival after 1922 of masses of refugees from Asia Minor leads to a hectic planning effort to create new settlements. Planning is no longer a mere matter of approving plan amendments.

8.2 Historical Events and Legislation

The autonomous state of Crete had its own constitution, with the backing of the Great Powers, its own High Commissioner, Prince George, son of King George I of Greece, and its own government, of which Eleftherios Venizelos is minister of justice. However, a new revolt erupted in 1905 claiming the right to join Greece (Vakalopoulos 2005, 299–306). Three years later the Ottoman Empire reaches its demise after the insurrection of the Young Turks, under Mustafa Kemal Atatürk (Veremis and Koliopoulos 2013; Beaton 2020). The situation in Greece is in flux. King George survives an assassination attempt in 1898, but Deliggianis, former prime minister, is killed in 1905 in front of the Parliament building (Beaton 2020, 171–172). A slow, gradual process of outmigration to America begins, with most serious long-run repercussions. About a quarter million rural workers emigrate, at a time when 65% of the working population was employed in the agricultural sector (Veremis and Koliopoulos 2013, 181; Dertilis 2019, 205). In the period 1890–2020, the net inflow of migration to the United States reached 18.2 million (Fernández-Armesto 2002, 389). In Greece, at a time of an effort to industrialize and strengthen the urban economy, we witness a replacement of internal rural migration to the cities by a massive trans-Atlantic exodus. In the words of Panagiotopoulos (1985, 523), the Greek city was unable to attract the nascent proletariat and witnessed it change its migratory course and migrate to America. These developments inhibited industrial growth, in spite of legislation enacted in 1903 and 1910 in support of industries (Anastasopoulos 1947, vol. II, 787), the foundation of the Union of Industrialists and Craftsmen in 1907, and the changing profile of the Greek industrialists, of whom many had been educated in Germany and Switzerland (Agriantoni 2021).

However, industrial production continued to lack specialization and to be associated with commercial activities. There was an absence of peak industries able to alter the course of the Greek economy, a role which was rather assumed by the merchant shipping sector (Tsoulfidis 2015, 212–213). The ascent to power in 1910 of the Liberals under Venizelos, following a military uprising in 1909, could potentially change the profile of the economy, if the rest of the 1910s were not to be a period of continuous wars, dramatic domestic political divisions, and ultimately, in 1922 of a disastrous military campaign. Still, the Liberals changed the national Constitution in 1911 and in two years voted 337 new laws (Beaton 2020, 189–190), including a law which created a new ministry of Public Works, with town planning powers. The new constitution opened the way to compulsory land acquisition for the public benefit. The term "benefit" replaced the term "need" of the previous constitution (Alivizatos 2011, 163–166; Kaltsonis 2017, 84–85). Land granting to landless peasants or to refugees—this time from Bulgaria and Romania—a leftover from the past, was one the reasons.

The importance of the provisions of the 1911 Constitution was great from a spatial planning perspective. The compulsory purchase of land for reasons of public benefit, the compensation paid to erstwhile owners, and the arbitration role of law courts, became a standard process till the present. Of course, the problem of the protracted compensation procedures, when land is expropriated for common use, has not disappeared. Other variables gained prominence since then, like land use and environmental protection, but this does not diminish the innovations of the 1911 Constitution. As far as the distribution of land to various social groups in Thessaly is concerned, we have a repetition of the fragmentation of land, which underlies the later widespread out-of-plan construction.

A new law was voted in 1912 on municipalities and communes, which replaced a law of 1833. It bears the signature of King George I, who was assassinated a year later (Beaton 2020, 195). The period is one of wars and domestic conflicts: the Balkan Wars of 1912–1913, the liberation of northern regions, the split between Liberals and Royalists which results in two antagonistic south and north governments, and the entry of Greece in the Great War on the side of the *Triple Entente* allies, after the latter had blockaded the port of Piraeus. Italy had in the meantime occupied the Dodecanese islands, which remained under Italian rule until 1947 (Vakalopoulos 2005, 365–390, 401, 410). As throughout the nineteenth century, the traditional Protecting Powers (England, France, Russia), now as members of the *Entente Cordiale*, had once again played a key role in the decisions of Greece, this time in the war against Germany and Austria-Hungary, which had been joined by the Ottoman Empire and Bulgaria, i.e., the defeated forces of the Balkan Wars (Roberts 1996, 448). For Greece, World War I was a quasi-prolongation of the Balkan Wars and was in fact extended even further with the military campaign in Asia Minor, which resulted in defeat, the destruction of Izmir, the expulsion of the Greek minority and the gigantic flow of refugees. All these dramatic developments changed repeatedly the spatial structure of the country and its population composition, and caused an urgent need to accommodate the refugees.

National unity, social cohesion, and political normality had been shattered (Alivizatos 2015, 77). Town planning too had to respond to this enormous challenge. It also had to absorb new planning ideas and institutional arrangements in the European environment.

8.3 Great Britain, France and Germany

Taking his cue from the impact of the garden city ideas, Benjamin Ward discussed the theories of innovation dissemination. He took as an example the spread of British and French concepts and practices, of which the town planners and *urbanistes* themselves were the carriers, such as the Frenchmen Henri Prost, Alfred Donat Agache και Ernest Hébrard. The last one was involved as chief planner in the reconstruction of Thessaloniki, after the 1917 wildfire. As sources of knowledge transfer, Ward mentions German planning of urban extensions and the French housing estates. As others before him, he adds that the British considered Germany as the origin of planning ideas and as an example of effective and systematic planning, although with some dark shades of an authoritarian, obstinate and ruthless approach (Ward 2018, 76–87). In England, statutory town planning is in fact introduced by the Town Planning Act of 1909, later revised in 1919 (Heap 1969. 5–7). Municipalities were now obliged to produce plans to secure sanitary conditions, amenity, and convenience, but the difficulties facing small municipalities caused an amendment limiting the obligation to municipalities with more than 20,000 inhabitants. The revised law was now the Town and Country Planning Act, which required that the plans should be extended to land which was not destined for building. 1914 was the year when the Town Planning Institute (TPI) was founded, the professional association of town planners, renamed in 1959 to Royal Town Planning Institute or RTPI (Cherry 1996, 38). What in effect happened after the 1909 act was that professionals of a diverse background (surveyors, civil engineers, architects, jurists) began to work together in local government town planning departments and realized that they shared a common professional activity. A cross-professional conference held in 1913 resulted in the creation of TPI, with Thomas Adams as its president, with the aim to promote the study of town planning, civic design, and urban development, and to care for the new profession. Simultaneously with the English 1909 act, a book appeared in the USA by Benjamin C. Marsh, entitled *An Introduction to City Planning* (Logan 1976, 377). In 1916 the zoning regulation of New York is introduced (Ladd 1990, 9–10), and in 1917 the American City Planning Institute is founded (Ward 2018, 79).

It was in this creative context, that in 1915 the pioneering work of the Scottish biologist Patrick Geddes appeared, under the title *Cities in Evolution.* Geddes was the first to formulate the sequence "survey-analysis-plan", which in due course became a rule, and the triptych "place-work-folk" (Geddes, 1968; Wassenhoven 2024a). His contribution is elaborately explained by Meller (1981, 46–71), who stressed that Geddes was anxious to discover the relationship between social processes and spatial forms. Geddes considered the collection of reliable data as a fundamental step in the planning process. His "biological" methodology was successfully tested in India. His contribution was one more valuable tile in the mosaic of precious theoretical writings which enriched town planning. Although this was a very troubled period, messages travel across the English Channel and through the European Continent reach distant lands. The messengers are the lucky ones who study or work in the countries of the messages' origin, or simply visit them in search of information, which thus becomes known in a place like Greece. What is impressive is that before the English and French planning laws, a path-breaking planning law had been adopted in the Cretan State, which undoubtedly influenced later initiatives in Greece.

The exchange of ideas was visible in France, in both town planning professional and theoretical matters. It was of course from England that the garden city ideology was exported to France through the writings of Georges Benoit-Lévy (Phalip 2019, 73). The books by Josef Stübben and Camillo Sitte were made available in French translations at the beginning of the twentieth century. The translation of a book by Raymond Unwin followed. Prophets of the new planning science in France were the architects Tony Garnier and Eugène Hénard, the engineer J.C.N. Forestier, the economist Benoit-Lévy and P. de Souza. However, the first manual in the French language was that of the Belgian L. Van der Swaelmen, published in 1916. Shortly afterwards appear the book of by Auburtin and Blanchard and that by D. A. Agache (Gohier 1965). In 1912, the first public offices of financially accessible housing are established in the Seine district, on the initiative of Henri Sellier (Phalip 2019, 74), who in 1918 founded, with the famous writer Marcel Poëte, a planning school for advanced studies and started a journal on cities. Planning legislation is at the same time reinforced, as various decrees are published on noxious, insanitary and dangerous activities (Rossillion 1963, 204–205). Of particular interest for Greece is legislation on land readjustment. The so-called Chauveau law of 1918 asserted the value of land-reparcelling (*remembrement*) (Rougean and Sagarminaga 1994).

Many writers consider the 1919 *Loi Cornudet* as the first French town planning law, which was later revised in 1924. It has been very popular among Greek writers, as indeed was the German *Lex Adickes*, which I mentioned in Chap. 7 (Kafkoula 2007, 112; Karadimou-Yerolympos 1995, 41). From a Greek perspective, its importance lies in the fact that it precedes by four years only the Greek legislative decree on towns, townships and settlements, of which the writers were no doubt aware of French legislation and of the 1901 act of the Cretan State. As a consequence, the Cornudet law deserves some additional comments about its importance, but also its failures.

Until then, there were no comprehensive legal instruments to support town planning in France. There existed only disjointed tools to regulate road tracing, public spaces, sanitation, heritage protection, and dangerous installations. The needs of reconstruction and expansion of cities after World War I demonstrated the inadequacy of existing legislation. Already in 1909 proposals of town planning laws had been submitted to parliament, similar to those existing in Sweden, Holland and Great Britain (Jacquot et al. 2019, 32–33). The law introduced in 1919 by Honoré Cornudet des Chaumettes, a member of parliament, imposed the production of plans for towns over 10,000 inhabitants, but after protracted delays only a quarter of municipalities were able to comply, although the law was revised in 1924. It was only in 1943 that a system of control was installed to oversee plan production. (Prévost et al. 2012; Orillard 2018; Davis 1966; Gohier 1965; Kalflèche 2018). The Cornudet law was the result of the innovations proposed by the so-called Social Museum, which included among its members distinguished architect-planners (Ghorayeb 2019), who placed town planning at the juncture of science and art, even philosophy. One of them was Henri Prost, who, in 1913–1914, worked in Morocco, at the invitation of the governor Lyautey, designed cities, and drafted planning legislation, in cooperation with Guillaume de Tarde. It is an irony that for a while a colony had more advanced planning legislation than the colonial metropolis. After the Cornudet law, planning science had at its disposal operational tools like the *Plan d'Aménagement, d'Embellissement et d'Extension* or PAEE for curing the ills of the industrial city. However, as Gohier points out, the law proved to be full of gaps and planning could not be applied narrowly only at the level of municipalities, without planning at a metropolitan and regional level. In the aftermath of the war, conditions did not favour slow planning processes. No one was prepared to exploit the opportunity of reconstruction to change the organization of cities. Municipalities faced difficult economic circumstances and the availability of experienced planners was limited. Out of 2300 towns which had to produce a new plan, only about 600 managed to do so until 1939. The PAEE plan had to be based on a plan of street alignment and ground levelling (*alignement et nivellement*) and it was the only plan on which granting of building permissions was to be based. The French building permission (*permis de construire*) was at the same time a planning permission (Albertini et al. , 1993, 336). Today, however, a planning certificate (*certificat d' urbanisme*) is usually required before the building permission.

With regard to Germany, I have already mentioned the building regulation of Saxony of 1900, which touched the sensitive issue of private property rights. The main body of German *in rem* property legislation was codified in the Civil Code of the same year, but was transferred to a 1919 regulation of estate rental (Hertel and Wicke 2004). In this context, Frankfurt followed a systematic land policy to eliminate speculation and impose a tax on unearned surplus value (Ladd 1990, 195). Frankfurt, Ulm and Freiburg also pursued a policy of land acquisition, and by 1906 Frankfurt and Ulm owned half of the land in their jurisdiction, with Freiburg owning more than three quarters. Frankfurt also readjusted the traditional methods of long-term leases, possibly copying a British equivalent policy (*op.cit.*, 198–199). A 1918 Prussian housing law improved prior design regulations, but was not yet a comprehensive building regulation (Pahl-Weber and Henckel 2008). In Germany, as in England and France planning thinking is consolidated, the first planning journal appears in 1906, the first town planning chairs are founded in universities, and the first academic programmes are introduced (Albers 1986).

8.4 Planning in Greece

There have been many influential personalities, politicians and planners, with a key role in Greece, who were certainly exposed to these developments, like the planners Anargyros Demetrakopoulos and Ernest Hébrard, the minister responsible for planning Alexandros Papanastasiou, even the Prime Minister Venizelos. Their role in the reconstruction of Thessaloniki and then in the writing of the 1923 planning decree was decisive. The 1923 legislative decree on town planning was a major innovation (Melissas 2024; Wassenhoven 2024b). The fact that politicians and planners were able to continue planning reforms after the trauma of World War I, is impressive. For Greece, the Great War was in essence the third Balkan War (Roberts 1996, 448). The cost of the war to Europe in human lives has never been estimated precisely but it reached ten million deaths, without counting the losses caused by disease (*op.cit.*, 455; Fernández-Armesto 2002, 471). Yet, town planning entered a new phase, under the influence of the Fordist model and of modernist ideas (Schröteler-von Brandt 2018).

Before describing planning activity in the state of Greece, a short reference is in order to the 1901 decree of the Cretan State on the plans of towns, townships and villages. According to Trimandili-McGann (2005, 402–409), the minister of justice Venizelos must have been involved in its preparation, with an active contribution by the French officer G. Durand. Plans for the cities of Chania and Herakleion were then prepared (Kladou-Bletsa 2005, 413–414). The decree, which is available in the Greek State Archives, included thirteen chapters. The first chapter is essentially a repetition of the 1835 Greek Royal Decree. The third chapter focuses on the perimetrical free zones of towns. The remainder is in fact a building regulation, without any reference to land uses, common spaces, central functions, zoning, public participation, property readjustment, and new settlements. Nevertheless, it was an important precedent, if only for the role of Venizelos, who a decade later became Prime Minister of Greece.

In Greece, certain laws and decrees of the first twenty years of the twentieth century are worth mentioning, for example a 1909 law on road building, because of its reference to expropriation and compensation, and a 1910 law on the responsibilities of the Public Works Department, which include the rejuvenation and *kallopismos* (embellishment) of towns and townships. The use of terminology often betrays the origin of concepts. The term *kallopismos* is no doubt a translation of the French term *embellissement*, first used in France in an 1852 law (Davis, 1966). A 1914 law transferred town planning duties to the new Ministry of Transport, which was soon to play a crucial role in the reconstruction of Thessaloniki after the 1917 fire and for the drafting of the major town planning act of 1923.

In 1920, the gate tariffs, which I mentioned in Chap. 6, were cancelled by law, and a royal decree on the maximum height and other features of buildings, was signed by King Alexander, who was soon to die after being bitten by a monkey at the royal country estate of Tatoi. The decree was made necessary by the increased building activity in urban centres, especially in Athens, of which the population more than doubled in about 20 years (Monioudi-Gavala 2017, 24; Demetrakopoulos, 1934–1937, 408). Dozens of amendments were made to the Athens plan and the capital was divided into new neighbourhoods (Monioudi-Gavala 2017, 88–99). According to a 1909 law, when an owner loses the whole or part of his property after the implementation of a plan, he is compensated by his neighbours. For the implementation of the plan the prior opinion of the municipal council is now required. The political and military adventures of the period did not prevent the approval of a large number of town plans, among which one can discern suburbs of Athens, villages of Attica, and towns of Macedonia and Crete, now part of the Greek state (Demetrakopoulos 1934–1937; Vakalopoulos 1988).

At a level above the individual town plan amendments, efforts started to produce an Athens metropolitan plan, and proposals were submitted by the German Ludwig Hoffmann in 1908 and the English Thomas Hayton Mawson in 1914, in the latter case at the request of Queen Sophia, proposals which were never realized (Monioudi-Gavala 2017, 23; Demetrakopoulos 1934–1937, 421; Biris 2005, 277–278). A plan by the engineer Balanos was practically ignored. One more effort was that of 1918 of a committee under Stylianos Leloudas, which apparently was the first to extend its interest to the entire Athens basin and to take into account all the legal and economic implications (Biris 2005, 280). Yet, in 1919, the Minister of Transport Papanastasiou invited another foreign consultant, the experienced Frenchman Ernest Hébrard, to try his hand. Germany, England and France had by now secured their presence. The efforts continued in 1920 by a committee headed by Petros Kalligas, and with the participation of Demetrakopoulos. Their plan had a similar fate. It was shelved in 1924 at the time of a brief military dictatorship (Karadimou-Yerolympos 2012; Kafkoula 2007, 249; Stamatiou 2005), probably because of the opposition of affected land owners (Demetrakopoulos, 1934–1937, 421). It seems obvious that in spite of the realization that a more strategic approach was required, the repeated attempts to produce a comprehensive strategic plan failed miserably, including that by Hébrard, who, without doubt, was aware of developments in France and of the first draft of the law Cornudet, which had been submitted to the French parliament as early as 1912. In an article on the work of Hébrard in Athens, based on his own diary, Kafkoula and Karadimou-Yerolympos (2008), confirm his thesis for a global approach to the Athens conurbation (Fig. 8.1).

Before his involvement in Athens, Ernest Hébrard had been the chief planner in the reconstruction of newly liberated Thessaloniki, after the 1917 fire, a task described in detail by Karadimou-Yerolympos (1995, 2016) and Demetrakopoulos (1934–1937). Several laws and decrees were issued to enable this reconstruction. The tool of "property groupings", a variant of land readjustment, was extensively used for the implementation of the reconstruction. The planning team functioned from the start with Hébrard, who was in Macedonia as director of the Archaeological Service of the French army, during the Great War, his French assistants, Thomas Mawson, and several Greek architects and engineers. The Thessaloniki experience had great influence on the town planning legislative decree of 1923. It was described by the French town planning historian Pierre Lavedan as the first great planning project in Europe in the twentieth century (Karadimou-Yerolympos 1995, 35).

Fig. 8.1 The destroyed area of Thessaloniki. *Source* Karadimou-Yerolympos (1995, 78). Available in Album 'Incendie de Salonique. 18–19 Août 1917' of the Aeronautical Service of the French Army/ Thessaloniki Fire 1917 Map

Thessaloniki was not the first of the Greek new Macedonian cities to be ruined by fire. Serres was burned down by the receding Bulgarian forces during the 2nd Balkan war (Karadimou-Yerolympos 1995, 72–74). The reconstruction of Serres was in a sense a first rehearsal of the intervention in Thessaloniki. These painful experiences were the opportunity for the introduction of planning innovations in Greece. The institutional edifice of Greek planning was enriched with impressive speed. However, the arrival of 1.3 million Asia Minor refugees after 1922 was to cause a new major challenge.

8.5 Conclusion

The main components of the period discussed in this chapter were, from the Greek perspective, the reconstruction of Thessaloniki after the fire of 1917, and from an international view, the progress on planning theory and practice. Of relative importance was also the 1901 planning law of the Cretan State, which had an impact on the Greek legislative decree of 1923. The attempts to plan the Athens conurbation in a comprehensive way proved vain, although they demonstrate a rising consciousness that urban planning has to adopt a more strategic orientation. The Thessaloniki reconstruction project added important tools to the Greek planning arsenal. It demonstrated, once again, the continuing process of transfer of concepts from other European countries, mediated through the work of experienced foreign planners. This was a process which continued later during the settlement of refugees.

In this chapter, attention was paid to planning ideas and reforms in England, France and Germany, which much later found their way to Greece. With regard to England, emphasis was given to the rise of the professional status of town planners and to the conceptual contributions of Ebenezer Howard and Patrick Geddes. Howard's effect on Greece is limited to fashionable references in the planning literature and to garden city imitations in suburbs of Athens. Geddes however introduces the methodology of survey-analysis-plan long before the advent of the rational systems approach, which, in the long run, was due to influence the structure of spatial plans, even in the present Greek planning system. As to France, this chapter dealt with legislative innovations, mainly the Loi Cornudet, which was a great hit in Greek literature, and with the establishment of terms like *urbanisme* and *urbaniste*. A similar trend was dominant in Germany, although *Städtebau* had been adopted earlier. One is however impressed by the circulation of ideas between countries.

As in previous chapters, Greek planning legislation was reviewed along with the production and amendment of plans, which evolve into a routine to the satisfaction of vested interests. Despite the criticism, the conditions outlined in this chapter are the introduction to major reforms, which were sadly to be interrupted by the Asia Minor disaster and the ensuing refugee problem of the interwar years.

References[1]

Agriantoni C (2021) German technological superiority in Greece, 1900–1940: practices and conceptions. In: Kyrtsis AA, Pechlivanos M (eds) Compendium of Greek–German encounters. URL: https://comdeg.eu/el/compendium/essay/104279/ [GR]

Albers G (1986) Changes in German town planning: a review of the last sixty years. Town Planning Review 57(1):17–34

Albertini J-B, Berenguer C, Marx J-L (1993) Dictionnnaire juridique: Pouvoirs locaux. Dalloz, Paris

Alivizatos N (2011) The constitution and its enemies in modern Greek history, 1800–2010. Polis, Athens [GR]

Alivizatos N (2015) Pragmatists, demagogues, and dreamers: politicians, intellectuals and the challenge of power. Polis, Athens [GR]

Anastassopoulos GA (1947) History of Greek industry, 1840–1940. 3 volumes. Elliniki Ekdotiki Etaireia, Athens [GR]

Beaton R (2020) Greece: Biography of a modern nation. Penguin Random House, London

Biris K (2005) Athens: from the nineteenth to the twentieth century, Revised edition. Melissa, Athens [GR]

Cherry G (1996) Town planning in Britain since 1900: the rise and fall of the planning ideal. Blackwell, Oxford

Davis J (1966) Zones à Urbaniser en Priorité et Zones d' Aménagement Différé. Conférence. Ministère de l'Équipement, Paris

Demetrakopoulos A (1934–1937) Town plans. Techniki Epetiris tis Ellados, 1934–1937, vol B (II), pp 359–449 [GR]

Dertilis G (2019) History of modern and contemporary Greece, 1750–2015, Revised edition. Crete University Press, Herakleion [GR]

[1] References in Greek are marked with the initials GR.

Fernández-Armesto F (2002) Civilizations. Touchstone—Simon & Schuster, New York
Geddes P (1968) Cities in evolution, 1st edition 1915. Ernest Benn, London
Ghorayeb M (2019) La loi Cornudet, un urbanisme hygiéniste et social. Droit et Ville, 88(2):43–58
Gohier J (1965) L'évolution de l'urbanisme en France. CRU65—Études et Essais. Centre de Recherches d'Urbanisme, Paris
Heap D (1969) An outline of planning law, 5th edn. Sweet & Maxwell, London
Hertel C, Wicke H (2004) Real property law and procedure in the European Union/National Report: Germany. European University Institute (EUI) Florence/European Private Law Forum, in cooperation with Deutsches Notarinstitut (DNotI), Würzburg
Jacquot H, Priet F, Soazic M (2019) Droit de l'urbanisme. Dalloz, Paris
Kafkoula K (2007) The adventure of garden cities. University Studio Press, Thessaloniki [GR]
Kafkoula K, Karadimou-Yerolympos A (2008) Ernest Hébrard and the plan for Athens (translation from French original) [GR]. https://www.academia.edu/23127489/
Kalflèche G (2018) Droit de l'urbanisme. Thémis Droit/Presses Universitaires de France, Paris
Kaltsonis D (2017) Constitutional history of Greece, 1821–2001. ΚΨΜ, Athens [GR]
Karadimou-Yerolympos A (1995) The reconstruction of Thessaloniki after the 1917 fire. University Studio Press, Thessaloniki [GR]
Karadimou-Yerolympos A (2012) Anargyros Demetrakopoulos: town planning as ideology. In: 170 years of the Polytechnic, engineers and technology in Greece. Conference proceedings, vol II. National Technical University of Athens, pp 149–169 [GR]
Karadimou-Yerolympos A (2016) Fortified military camp, defenseless city: disaster and transformation of Thessaloniki after the 1917 fire. University Studio Press, Thessaloniki [GR]
Kladou-Bletsa A (2005) The city of Chania and Eleftherios Venizelos. In: Eleftherios Venizelos and the Greek city: town planning policies and socio-political re-arrangements. Conference proceedings, Chania, 24–27 Oct 2002. National Foundation of Research and Studies "Eleftherios Venizelos", Technical Chamber of Greece and National Technical University of Athens, Athens, pp 411–424 [GR]
Ladd B (1990) Urban planning and civic order in Germany: 1860–1914. Harvard University Press, Cambridge (Mass)
Logan Th-H (1976) The Americanization of German zoning. J Am Inst Planners 42(4):377–385
Melissas D (2024) The legislative decree of 1923: a turning point in Greek town planning. In: Serraos K (ed) The legislative decree of July 17th, 1923, 100 years later. Sakkoulas, Athens, pp 125–139 [GR]
Meller H (1981) Patrick Geddes 1854–1932. In: Cherry GE (ed) Pioneers in British planning. The Architectural Press, London, pp 46–71
Monioudi-Gavala D (2017) Planning and real estate property in Athens, 1833–1922. Paraskinio, Athens [GR]
Orillard C (2018) Urbanisme and the Francophone sphere. In: Hein C (ed) The Routledge handbook of planning history. Routledge, London, pp 161–172
Pahl-Weber E, Henckel D (eds) (2008) The planning system and planning terms in Germany: a glossary. Academy for Spatial Research and Planning (ARL), Hanover
Panagiotopoulos V (1985) Rural exodus and labour force constitution in the Greek city. In: SSNH/Society for the study of New Hellenism (1985b), The Neo-Hellenic city: Ottoman heritage and Greek state. Conference proceedings, vol II. Athens, 26–28 Sept 1984, and Ermoupolis, 29–30 Sept 1984. Athens, pp 521–531 [GR]
Phalip L (2019) Histoire de l'urbanisme. Bréal, Paris
Prévost A et al (2012) The urban planning of French cities and the challenge of sustainable town planning: Improvement and limits. In: AESOP 26th Annual Congress, July 2012, Ankara. HAL archives-ouvertes. hal-01179362
Roberts JM (1996) A history of Europe. Helicon, Oxford
Rossillion P (1963) Les plans d'urbanisme. Berger-Levrault, Paris

Rougean P, Sagarminaga Y (1994). Remembrement et aménagement foncier en France. D.E.S.S. "Aménagement integré des territoires". Institut National Agronomique de Paris-Grignon 17:81–113

Schröteler-von Brandt H (2018) The history of urban planning. Academy for Territorial Development in the Leibniz Association (ARL). https://nbn-resolving.org/urn:nbn:de:0156-55997441

Stamatiou E (2005) The policy of the Ministry of transport for the interwar city: socio-economic framework and modernizing directions. In: Eleftherios Venizelos and the Greek city: town planning policies and socio-political re-arrangements. Conference proceedings, Chania, 24–27 Oct 2002. Athens: National Foundation of Research and Studies "Eleftherios Venizelos", Technical Chamber of Greece and National Technical University of Athens, Athens, pp 223–235 [GR]

Trimandili-McGann T (2005) Cretan State: years of autonomy, administrative organization, and the legislative work of Eleftherios Venizelos for the town plans of Crete. In: Eleftherios Venizelos and the Greek city: town planning policies and socio-political re-arrangements. Conference proceedings, Chania, 24–27 Oct 2002. Athens: National Foundation of Research and Studies "Eleftherios Venizelos", Technical Chamber of Greece and National Technical University of Athens, Athens, pp 401–410 [GR]

Tsoulfidis L (2015) Economic history of Greece. University of Macedonia Press, Thessaloniki [GR]

Vakalopoulos A (1988) Single-volume history of Macedonia. Kyriakidis Bros, Thessaloniki [GR]

Vakalopoulos KA (2005) History of Greece/Single volume synthesis: from the genesis of new Hellenism (1204) to modern Greece. Ant. Stamoulis, Thessaloniki [GR]

Veremis ThM, Koliopoulos GS (2013). Modern Greece: a history from 1821. Patakis, Athens [GR]

Ward SV (2018) Planning diffusion: agents, mechanisms, networks and theories. In: Hein C (ed) The Routledge handbook of planning history. Routledge, London, pp 76–90

Wassenhoven L (2024a) Theories and praxis of spatial planning: from the invisible to the visible. Propobos, Athens [GR]

Wassenhoven L (2024b) The 1923 legislative decree: historical conjuncture, comparisons, innovations, and the problem of out-of-plan construction. In: Serraos K (ed) The legislative decree of July 17th, 1923, 100 years later. Sakkoulas, Athens, pp 19–36 [GR]

Chapter 9
General Conclusions

Abstract Chapter 9 is devoted to general conclusions. It is an attempt to summarize the previous chapters and draw lessons from the town planning path of Greece during the years from the naval battle of Navarino of 1827 to the legislative reform of 1923.

9.1 What Does It Mean to Be a Dependent State?

My entire argument on the development of town planning in 19th century Greece has been closely related to the effective subjection of the country to the control of its Protecting Powers and to the constant flow of imported ideas from the developed European North and West. Greece was in a permanent condition of financial indebtedness and a never-ending reliance on the guarantees of the great powers in contracting loans in the international markets. On several occasions it was openly blackmailed by the blockading of its ports by foreign naval forces. Its very existence as an independent state and its subsequent territorial expansion were due to English, French and Russian military support or the diplomatic maneuvers of these powers, vis-à-vis the notorious Eastern Question. Its spatial structure had an immediate impact on the growth of its cities, the settlement network and the potential of urban centres, i.e. on factors town planning had to deal with. This was a classical case of dependence. I have discussed the question of dependence before (Wassenhoven 2022), therefore I hope I will be excused if I repeat myself.

The Greek economy and the country's political stability were totally dependent on the support and on the economic interests of the Protecting Powers. The transfer of a term like "dependence" from the condition of post-Second World War developing countries to that of Greece in the 19th century may seem too bold but, in my view, justified: "By dependence", Dos Santos suggested, "we mean a situation in which the economy of certain countries is conditioned by the development and expansion of another economy to which the former is subjected" (Dos Santos 1973, 109). Dependence, in his view, is a "conditioning situation": "Regarding concepts that try to explain peripheries in terms of political power, there is the long tradition of 'Dependency theory'. This is a school of thought that attempts to explain the

L. C. Wassenhoven, *The History of Town Planning in a Dependent State*,
https://doi.org/10.1007/978-3-031-96958-4_9

'underdevelopment' of regions and countries within the capitalist world economy. Emerging in Latin America during the 1960s, these theories have argued that there is a power asymmetry between developed and underdeveloped states. 'Underdevelopment' is explained with reference to the neo-colonial 'domination' of the periphery by capitalistic metropolis". Dependence in the case of Greece did not involve colonial domination, but was a permanent condition, even though it was not accompanied by military occupation (Sotiropoulos 2019, 41–42). The country was however constantly under the threat of gunboat diplomacy, a reality which reached a peak during the First World War.

After the 1893 bankruptcy, Greece had to suffer the humiliating imposition of foreign control, exercised by an International Financial Commission to which half of the state's revenues were directly assigned. The plague of public finances continued to force the state into dependence on foreign and expatriate capital, with little effect for industrial development. Foreign imperialism played a major part in the rise of the bourgeoisie during the last years of the 19th century, through the flourishing of what Svoronos (1972, 60) describes as "peripheral Hellenisms", the Greek expatriates of Russia and the Near East, who dominated the economy of the new state and conferred to it an importance which it could not otherwise acquire, given its small size (Markezinis 1966, vol.I, 295). Big international lenders and public loan managers, Dertilis (2019, 460) observes, were in a position to always adjust in time their investment strategy on the basis of sound predictions, which was simply impossible for simple bond holders in Greece. According to Svoronos, it was Istanbul and not Athens which was the economic capital of the Greeks, although Athens became their cultural centre. Dependence on external centres and domestic resistance to it are frequent themes in historical studies, what Sotiropoulos (2019, 50) calls *Svoronean* historiography, but it is no doubt true that the territory of Greece, not for the first time its history, had to rely on external space. As a theoretical approach, the use of dependency theory, can now be classified in the broader spectrum of Southern theory, of which "a central theme is that ideas deployed in much social theory and description originate in the north of the world and that ideas originating in the south are ignored in these hegemonic accounts" (Mabin 2017, 23).

9.2 Why a Return to the Past?

Is the past a heavy burden on our shoulders? Does it have lessons to teach us? Is it admirable, terrifying or indifferent? Should we remember it or is it better to forget it? Did it have moments for which we should be proud of, or also memories we should push under the carpet? The interest in this past is only a matter for historians? And what historians? Those who curse it or those who deify it? Should it concern all of us? Is this past teaching us who we are and why? Does it explain our achievements but also our miseries? Does it provide an interpretation of the bleak environment where we live? Does it tell us what we had and lost, or what we were rescued from and what we gained?

Are these questions relevant when we ask them with respect to the "towns and townships" where we live, to recall the title of the Greek Royal Decree of 1835? Or also to the "villages" included in the 1901 law of the Cretan State? Actually, do we still remember the pleasant features of our city, its countryside, our neighbourhood, or our village, which have been effaced forever? This is the reason why a return to the past is so important. To understand what was erased, demolished, created and how. Since we live, willingly or not, in a space of housing, work, production and leisure, which is described as town or settlement, is it not desirable to know how and through what trajectories it was created or transformed? Should we not search the ideas, actions, plans and initiatives which produced these transformations? The answer is yes, even though our findings may be depressing.

9.3 Searching the Meaning of Concepts

Truth is concealed in words and in their meaning. When an intention, decision or action is announced, the meanings embodied in them may be those of a genuine innovation, ambition, determination, ulterior motive, narrow interest, or procrastination. The meanings can be a genuine creative and inventive response to a dominant condition or conveniently borrowed from elsewhere, where they have been tried before. In town planning, when the craftsman-planner or the decision-maker have a vision for the future, their inspiration may have originated in the place where they are active or in the environment where they made their studies or worked in the past, or which they simply visited. Naturally, the pretence of inspiration may be simply the cover of a disguised material interest.

Meanings are not stable and they are not always expressed with the same technical or scientific terms. Over time, their content changes and is expressed by a new vocabulary, which later is considered obsolete. Alternatively, the terms embedded in this vocabulary, often sanctified by law, may either remain sacrosanct, or be adapted to changing circumstances. This means that the term does not necessarily disappear over time and may continue to shape our subconscious understanding. This is particularly true when the meaning of the term is inserted in legal rules which remain valid for several decades.

This is what happens particularly in town planning because it produces a solid, built environment which has a lasting life, but also because it takes place on the basis of legal rules which cannot be overturned overnight. In conformance with planning and building laws, administrative acts are issued for the approval of plans, a built environment is produced, buildings are erected, and public spaces are delimited, i.e., developments occur which are expected to last. A building permission or the amendment of a layout plan, which were approved under a legal regime which has been revised, do not cease to be legal. Ultimately, what we have to keep in mind is that the emergence, repetition, persistence or change of meanings is at the heart of the history of town planning.

9.4 What Was Town Planning?

This is not an easy question to answer, simply because thinking on the nature and role of town planning has been changing over time. Even if we limit ourselves to the 19th century and to Europe, we observe that the perceptions of the emerging class of town planners have not been stable. After a period when town planning meant no more than street tracing and fixing building location, definitions of town planning have been radically altered. I repeat here the definition given in 1904 by Goecke and Sitte in the journal *Der Städtebau*: "*Städtebau* is the unification of all technical and creative arts into a great enclosed entity … *Städtebau* is a science. *Städtebau* is a kind of art with specific aims of research, and specific big tasks for practical execution … in which numerous technicians, artists, national economists, hygenicists, social politicians, civil servants and lawyers are working together … with all inhabitants of our cities participating" (Kress 2018, 173). I also reiterate that in Britain, according to Cherry, the term "town planning" appears for the first time around 1906. The practice of conscious town building, according to which the constitutive elements of the town form are defined in accordance with deliberate design principles, is millennia-old and has been presented in a host of studies. Even without precise definitions in planning law, a new activity in urban affairs emerged in a short time-span, based on key ideas. Around this nucleus of concepts, Cherry concludes, a new corpus of knowledge was formed, which secured its presence in political and institutional structures (Cherry 1996, 17). Sanitary conditions, amenity, and convenience were now the key-words. As to France, as mentioned in an earlier chapter, the term *urbanisme* was first used at the end of the nineteenth century. It meant, according to Orillard (2018, 161), more than just town design. It incorporated a vast field of sciences, technologies and arts related to the structure and development of urban space, in order to enhance human wellbeing and social cohesion, while preserving the environment. Most francophone historians identify *urbanisme* with an "urban science".

What was the disguise of town planning in Greece, not just in legal texts such as the 1835 Royal Decree or in actual town plans, but also in social understanding? Officially, it was no doubt imitating the definitions of foreign planning systems. There were however special conditions of a social and cultural character, which made the adoption of alien practices extremely difficult. Real-life planning was in practice totally different from that prescribed in legislation and statutory plans. In spite of the transfusion of ideas from the West, their adoption was far from being real.

9.5 When Did Greek Town Planning Commence?

When the Governor Count Capodistria, described by some historians as authoritarian, despotic and populist, demonstrated his interest in the planned reconstruction of ruined towns and the creation of settlements for refugees, he starts from zero. The collaborators he recruited were mostly surveyors and army officers, in active service or in retirement. Their planning tools were street and square layout schemes, elementary division into housing quarters, location of public buildings, and building rules, such as the number of floors. We can also discern a search of conditions which would attract commercial activities, improve sanitation, and secure the consent of residents. We have here the first signs of environmental care, economic objectives and citizen participation. We also detect the presence of procedures, albeit at a primitive stage, of property exchange and land adjustment.

Independently of government policies, for reasons of custom-law, we also observe, when private properties are sold, the existence of pre-emption rights for the benefit of the nearest relatives, neighbours or simply fellow-citizens or for reverse priority rights when a property is sold back to the initial owner. A more active and open expression of state policy is the effort to create settlements for the reception of refugees. All these spatial planning initiatives are intensified in the Bavarian period when the 1835 planning decree is legislated and a series of fresh plans are produced, of which the plans for Athens and Piraeus are the best example. The ambitions of "Europeanization" and of the adoption of a monumental planning style are apparent. The protagonists of these efforts, Greek or foreign, are the product of an education received in foreign lands. The readers will hopefully excuse the author if he reminds them that the study of those years required copious research in bibliographical and archive sources.

9.6 Town Planning Between the Odyssean Clashing Rocks

From day zero, all renovating efforts aimed at the regeneration of towns are subjected to pressures from a broad variety of political forces, on the one hand of local chieftains, primates and kodjabashis, and, on the other, Greeks of the Diaspora, quickly named *autochthons* and *heterochtons*. The reason of this observation is not just theoretical or purely political. The internal struggles are quickly transformed into actions of economic character and of private interest. Town plans collide with land and property grabbing or with blunt refusals of plan implementation. The identities of politicians, ministers and large property owners frequently overlap. Although the behind-the-scenes role of the latter is of prime importance, planning is also undermined by small owners, whose interest is also in the hands of political leaders through networks of patronage. The evidence of these clientelist bonds is provided, throughout the nineteenth century, by the constant stream of amendments and extensions of town plans. Several of these interventions concern the high-quality urban districts and resorts where land belongs to high class owners, but a good many are located in the urban periphery, where the plans are hastily and carelessly extended. Town planning finds itself squeezed between the clashing rocks, the *Sympligades Petres*, like the vessel carrying Ulysses and his crew.

The clashing rocks are not only a matter of satisfying private interests, large or small. They also concern the shortage of public land, essential for the implementation of town plans. The problem was mainly present in Attica, the Athens region, where foreigners and locals had hastened to buy Turkish properties leaving no public land reserves. In general, the national lands, owned by the state after the independence war, soon became a puzzle, since they served as guarantee of loans contracted by the government. They were ultimately distributed to landless peasants in 1871, but the problem for town planning remained the lack of public land in small and large towns and the inability to provide sufficient land for common use or public facilities, unless private properties were compulsorily acquired and their owners compensated. The result was that town planning, caught between the clashing rocks of public debt servicing and private interest satisfaction, became decorative and limited to opportunistic and ill-thought arrangements.

9.7 Town Planning and Foreign Protectionism

For most of the 19th century, the Greek territory was stable, at least until the annexation of the Ionian Islands and Thessaly, which were in fact a present of the Protecting Powers. Until the end of the century, it was a country enclosed at the tip of the Balkan peninsula, with its northern frontier controlled by its former sovereign, the Ottoman Empire. Its maritime lanes were under the control of its protectors, who cared for their multiple interests in the Eastern Mediterranean.

Was this condition of interest to town planning? The answer is positive, because this asphyxiating environment determined the vigour of the country's urban centres, the prospects of industrial development, the attraction of human resources, and the needs the cities were supposed to satisfy. Weak urban centres, usually deprived of a creative social class with ambitions for their city, unable to support a regenerated urban fabric, and forced to survive in poor conditions, could not inspire bold planning innovations. This situation was mostly due to the fact that Greece was treated as a dependent statelet, a cushion between the advanced West and the Ottoman Empire, the so-called "Sick Man of Europe", a cushion the Protecting Powers considered important as passage to the Middle East.

A weak city, located in a country in perpetual transition, cannot handle a radical and ambitious town planning, able to make its citizens proud. It could be argued that the situation improves when the Eastern Question is near its end, in the closing years of the nineteenth century. But then, bankruptcy and military defeat function as a break, in spite of the first steps of industrial development and improvement of infrastructures during the premiership of Tricoupis. Athens is of course changing but is still nourishing an illusion of European grandeur. Transatlantic migration deprives of much-needed human resources. Town planning serves only to inflate the surplus value of small and large land properties. As uncontrolled urbanization floods the countryside and turns out-of-plan construction to a rule, town planning loses control.

9.8 The Ottoman Syndrome: *Efendi* and Client

Planning anarchy is convenient for all. Small owners are happy because their dream of a house in the middle of nowhere is realized. Large owners, who proceed to a subdivision of their estates to buildable parcels, enjoy welcome windfall profits. Political leaders, often large land owners themselves, perpetuate this situation with uncoordinated planning interventions, or turn a blind eye, by encouraging the opening of private and rural access roads and making promises of future public services. They retain the role of boss or *efendi* who takes care of his dependent citizen and political client.

I mentioned the ironical comment that after independence the local primate took over the role of the Ottoman aga. The problem is that now this syndrome is temporally extended, with the boss assuming a new identity, and the client rewarding the boss with his vote in democratic elections. Clientelism and patronage become the background of disparate planning regulations. Town planning is mostly limited to promote plan extensions, revise building conditions, and iron-out property anomalies with adjustments and parcel adaptations. National space is transformed to a huge terrain to be profitably exploited. The old non-buildable parcel is turned into a buildable plot. The town planner is reduced to a role of dealing with median walls between neighbouring properties and with building projections. The city, as the sum of urban spaces which secure human wellbeing and social interaction, while preserving the environment, is now reduced to an illusion.

9.9 My *Ospition* Above All

Spiti in everyday Greek language is the house, dwelling or just home. In the rather idiomatic language of the planning legislation of the 1830s, the term used is *ospition*, a medieval word from the Latin *hospitium*. For the average citizen, his *ospition* is what matters most. His hopes are usually defined as in the popular proverb, "better to have five in your hand, than to wait for ten"! It may be that his expectations change from a simple desire to give an *ospition* to his daughter as a dowry to the more complex objective of a higher surplus value of his house or to the acquisition of a second vacation home, but the family interest is always what drives his actions. If his immobile property is a vacant piece of land, his secret hope is that the appropriate planning office which issues building permissions will allow him to divide his land into two adequately-sized and buildable plots, which in due course will be integrated in a statutory town plan and be no longer dependent on an officially approved access road crossing his out-of-plan area. The "town planner" is the person who will help him escape from this embarrassing situation. For him, the town planner is neither the strategic programming official, nor the visionary designer.

All this may sound as an exaggeration, because a lot has changed since the independence war. However, the poor citizen of 1828 lived in an *ospition* in ruins and about to collapse. These conditions bred a mentality which withstood subsequent developments. Repeated economic crises and political instability perpetuated his mode of thinking and behaving. Admittedly, the security of owning a house, a perishable property in the legal jargon of those days, was slowly replaced partly by banking deposits and investment bonds, but the experience of fiscal crises and bankruptcies continued to keep the ownership of land and buildings at the top of private priorities. Owning a solid property remained the main guarantee of personal happiness and security, as even the economic crisis of the 2010 decade has demonstrated. The value of this property depends on the prescriptions and limitations imposed by town planning and building regulations. The owner therefore expects from the state to take the appropriate measures to increase his felicity and shield him against economic downturns.

9.10 The State and Me: A Love-Hate Relationship

Among several factors which determine the relations between state and citizen, town planning has a privileged place because it affects personal wealth. The general logic is that the citizen blames the state for anything that goes wrong, while expecting from it to solve all his problems. The state is responsible for everything, but is also the supreme protector who has to find answers for all our problems. In the town planning field, I may have broken the law, erected a building illegally, added an extra room without permission, trespassed on public land, encroached on a neighbouring property, bribed to obtain a building permission, and speculated in my real estate activity, but, ultimately, it was the state's failures and negligence which made me sin. The truth is that the state often overlooks illicit behaviour and encourages it indirectly, when, e.g., supplies electricity, water and transport infrastructure to illegal settlements or buildings. This practice started from the beginning of the new state's existence, with the exception of electricity. Even the first planning laws threaten illegal proprietors with demolition of their non-licensed structures, but then the authorities diligently search for solutions to legalize them, through plan extensions and use of ad hoc instruments, or by tolerating out-of-plan building activity. For years-on-end, the state turns a blind eye to irregularly urbanized areas, and then descends as a vulture to devour them, using all sorts of pretexts, including the protection of allegedly forested land and the irregular formation of low-density settlements. The citizen is then left open-mouthed and unable to understand why he is being persecuted.

The democratic citizen respects, in good faith, a sense of justice which must be heeded. The formal law sometimes disregards this faith, a problem which legal experts frequently acknowledge. When the state tolerates particular forms of behaviour for long periods of time, the citizen may rightly conclude that his behaviour does not break the law. In the first period of the new state, local tribunals respected traditional custom-law. It is natural to pose the question if the state was in a sense following a *sui generis* habit of tolerance to traditional illegal activities or whether it was breathlessly running after established practices. Prediction and prompt action were not among its strong points, certainly not in town planning. Successive governments attempted for years to plan the Athens conurbation in its entirety but failed. Town planning was stuck in trivial arrangements of building blocks, street alignments, building heights, and building coefficients.

9.11 Town Planning in the Hangman's Noose of Building Regulations

Town planning was in fact equated to a building regulation. In the 19th century, even this type of regulation was in fact an extension to all cities of the rules applying in Athens. This was a dogma which persevered for a century. Even after the town planning legislative decree of 1923, building regulations continued to be the gospel of planning, because of the absence of a cohesive planning code. The British Town Planning Act and the French *Loi Cornudet*, had no equivalent legislation in Greece. In the 1920s a succession of laws and decrees turned legislation into a real puzzle. Perhaps this was not an accident, but rather a reflection of the state's haphazard approach to problems. The turmoil of political events is no doubt an excuse, but the professional planner was swimming in a sea of uncertainties.

Town planning, as perceived at the outset, but even in the 20th century period before World War II, was a servant of building regulations, perhaps because it continued to be perceived as design of street layouts and individual buildings. It was the emergence of regional spatial planning, the rise of a rational, comprehensive form of planning, and environmental problems that were to change this attitude, although the problems did not disappear. As far as Greek planners are concerned, this state of affairs prevented for long the recognition of urban and regional planning as a separate profession and science, as in West European countries and the United States of America.

9.12 The Wisdom of Others

For all the periods discussed in this book, extensive sections were included describing developments in other European countries. The reason was the transfer of foreign practices and legislative experiences to Greece and their effect on Greek laws and plans. The application of alien theories, models, methods and instruments in the receiving territory is an issue which I could not possibly deal with in full. The transfer of knowledge from the North and East to the East and South has been recently the object of extensive theoretical studies, mainly related to the influence of developed and formerly colonial countries on less developed and dependent ones. The so-called Southern Theory is a prime example. Nevertheless, I would be wrong to ignore the fact that, at least in its first years, the Greek state lacked an institutional and organizational apparatus. The importation of foreign experience was a one-way street.

To trace the origins of imported innovation, the international literature had to be consulted, in order to form a picture of initiatives that gradually made their way into Greek practice. This study was arduous and time-consuming. However, it made possible an understanding of innovations beyond the confines of the Greek situation. It also brought to the surface concepts and terms which had to be singled out. The routes along which innovations travelled to reach Greece was a further major question.

9.13 Prophets and Knowledge Carriers

A real problem has been the identification of persons who helped the transfer of knowledge to Greece. It was not simply a matter of pinning down those who produced an idea or applied it in the real world. The practitioners and scholars, who introduced innovations and made a breakthrough are on the whole well-known and several writers have described their work and contribution. The difficulty is to identify the particular elements of their thinking which were copied and applied in a different context. It is even more difficult to find the carriers of this knowledge who brought it to Greece. They may have been individuals who worked in Greece and were physically present in the new state, or simply people who transmitted their knowledge by correspondence or through their teaching. This knowledge was not limited to the details of designing a town plan. It was also extended to side issues such as legal expertise, geodesical studies, sanitation and hygiene, or engineering. The mediators were sometimes in possession of knowledge which they had themselves promoted—e.g., in legal or archaeological scholarship—or people with the ability to transfer the knowledge of others and adapt it to local conditions.

The problems of identifying "prophets" and "knowledge carriers" were particularly acute in the early stages of the new state. In later years, particularly towards the end of the 19th century and the first decades of the 20th, conditions were more visible, the great innovators are widely discussed in the relevant literature, and the "carriers" were easier to identify. Town planning history as we enter the 20th century is richer and well documented. Events were better recorded, although, even as these lines are written, research continues and will probably continue to shed light on the fascinating story of town planning.

9.14 Manna from Heaven or a Hail of Stones?

At every stage of town planning developments in Greece, from the days of the national assemblies during the 1821 Independence War to the economic crises of the 1890s and 1930s and the continuous political upheavals, the uncertainty of unexpected events hangs over the heads of planning innovators and protagonists. On some occasions there was a feeling that the moment had come for radical change, on others the prevailing sense was that new initiatives had to be postponed indefinitely. There were short intervals which favoured a new start because normality had been restored, as in the time of Charilaos Tricoupis, or because of a wartime development which made possible a new beginning, e.g., when the northern frontiers of Greece were stabilized. There were also periods of introversion and decline.

In periods of great expectations, when the fall of "manna from heaven" is anxiously awaited, even though problems are still being accumulated, we witness legislative initiatives and town plan production with promising results. Such periods were those of the Capodistria government, the Bavarian Regency, the Tricoupis administration, and the first rule of the Liberals under Venizelos. On the other hand, there were times when the country was under a "hail of stones", when internal conflicts were in progress, when town planning is drawn into a routine of caring for petty-interests and of short-lived compromises.

9.15 Learning from Errors

The history of town planning in Greece in all the turbulent periods examined in this book has a lot to teach us. For the author it was an exciting adventure, as if he were a first-hand observer of unexpected changes of direction and traumatic experiences. He is convinced that town planners, no matter in what country they are currently active, have much to learn from this adventurous past and can benefit from the knowledge that the past has bestowed to them. This book is but a modest pebble in the mosaic of existing planning literature. A lot of the questions addressed in it deserve to be examined in greater depth, drawing from the example of many countries, developed or developing. I hope that this book will assist their future research.

References[1]

Cherry G (1996) Town planning in Britain since 1900: the rise and fall of the planning ideal. Blackwell, Oxford

Dertilis G (2019) History of modern and contemporary Greece, 1750–2015. Revised edition. Crete University Press, Herakleion [GR]

Dos Santos Th (1973) The structure of dependence. In: Wilber CK (ed) (1973) The political economy of development and underdevelopment. Random House, New York, pp 109–117

Kress C (2018) The German traditions of Städtebau and Stadtlandschaft and their diffusion through global exchange. In: Hein C (ed) The Routledge handbook of planning history. Routledge, London, pp 173–191

Mabin A (2017) Grounding southern city theory in time and place. In: Parnell S, Oldfield S (eds) The Routledge Handbook on cities of the Global South. Routledge, London, pp 21–36

[1] References in Greek are marked with the initials GR.

Markezinis S (1966) Political history of contemporary Greece, 1828–1964, 4 volumes. Papyros, Athens [GR]
Orillard C (2018) Urbanisme and the Francophone sphere. In: Hein C (ed) The Routledge handbook of planning history. Routledge, London, pp 161–172
Sotiropoulos DP (2019) Positions and contradictions of the Greek state in the 20th century, 1910–2001. Estia, Athens [GR]
Svoronos N (1972) Histoire de la Grèce moderne. Presses Universitaires de France, Paris (1ère édition 1953)
Wassenhoven L (2022) Compromise planning: a theoretical approach from a distant corner of Europe. Springer, Cham (Switzerland)

Additional bibliography for further consultation

Berry J, McGreal S (eds) (1995) European cities, planning systems and property markets. E & FN Spon, London
Catanese AJ, Snyder JC (eds) (1979) Introduction to urban planning. McGraw-Hill, New York
Cherry GE (ed) (1980) Shaping an urban world. Mansell, London
Cherry GE (ed) (1981) Pioneers in British planning. The Architectural Press, London
Cullingworth B (1997) Planning in the USA. Routledge, London
Dertilis G (2016) Seven wars, four civil wars, seven bankruptcies: 1821–2016. Polis, Athens [GR]
Drakoulis DP, Tsotsos GP (eds) (2014) Historical, social and town planning analysis of space: tribute to Prof. Evangelos Demetriadis. Antonios Stamoulis, Thessaloniki [GR]
Venizelos (2005) the Greek city: Town planning policies and socio-political re-arrangements. In: Conference proceedings, Chania, 24–27 Oct 2002. Athens: National Foundation of Research and Studies "Eleftherios Venizelos", Technical Chamber of Greece and National Technical University of Athens [GR]
Galantay EY (1975) New towns: antiquity to the present. Braziller, New York
Georgis G (ed) (2014) Governor Ioannis Capodistria: critical approaches and confirmations. Kastaniotis, Athens [GR]
Hein C (ed) (2018) The Routledge handbook of planning history. Routledge, London
Kontogiorgis GD (2020) Hellenism and Helladic state: two centuries of infighting, 1821–2021. Poiotita, Athens [GR]
Lagopoulos A-Ph (2017) Theory and methodology of town planning: from political economy to the semiotics of space. Patakis, Athens [GR]
Lüderssen K (1971) Planning and the legal system in Germany. Int & Comp Law Q 20(1):75–86
Ministry of National Education and Religious Affairs (1978) Year Capodistria—200 years from his birth: official speeches. National Printing Press, Athens [GR]
Mumford L (1966) The city in history. Penguin Books, Harmondsworth
Roberts JM (1995) The Penguin history of the world. Penguin Books, London
Serraos K (ed) (2024) The legislative decree of 17th July 1923, 100 years after. Sakkoulas, Athens [GR]
SSNH/Society for the study of new Hellenism (1985a) The neo-Hellenic city: Ottoman heritage and Greek state. In: Conference proceedings, Vol. I. Athens, 26–28 Sept 1984, and Ermoupolis, 29–30 Sept 1984. Athens [GR]
SSNH/Society for the study of new Hellenism (1985b) The neo-Hellenic city: Ottoman heritage and Greek state. In: Conference proceedings, Vol. II. Athens, 26–28 Sept 1984, and Ermoupolis, 29–30 Sept 1984. Athens [GR]
Svoronos N (1976) A review of the history of modern Greece. Themelio, Athens [GR]
Vergopoulos K (1975) The rural question in Greece: the social incorporation of agriculture, Exantas, Athens [GR]

Archival sources

Academy of Athens Archives
Aeon newspaper (1838–1888).
General Newspaper of Greece (*Geniki Efimeris tis Ellados*) (1825–1832)
General State Archives
Government Gazette (*Efimeris tis Kyverniseos*) (since 1832)
Greek Parliament Archives
National Newspaper (*Ethniki Efimeris*) (1832)

The Author: Curriculum Vitae

Louis C. Wassenhoven is Professor Emeritus, Department of Urban and Regional Planning, National Technical University of Athens (NTUA), member of the Association of Greek Urban and Regional Planners, member of DPU Associates [DPU: Development Planning Unit, University College London] and guest-editor of a special issue on planning in Greece of the journal Planning Practice and Research.

ORCID No: 0009-0004-4642-1203.

E-mail: lwassen@arch.ntua.gr or wassenhovenl@gmail.com.

The author holds a Master's degree in Architectural Engineering from the National Technical University of Athens, a Postgraduate Diploma in Planning from the Architectural Association School of Architecture, a Ph.D. degree from the London School of Economics and Political Science, and a Certificate of Planning Techniques of the French Government.

Past appointments, posts and memberships

- Full-time Senior Lecturer, Development Planning Unit, Bartlett School of Architecture and Planning, University College London;
- Visiting teacher, London School of Economics and Political Science;
- Member of the Royal Town Planning Institute;
- Visiting teacher, Bandung Institute of Technology (Indonesia);
- Professor, Department of Urban and Regional Planning, National Technical University of Athens;
- Director, Laboratory of Spatial Planning and Urban Development, National Technical University of Athens;
- Visiting Professor, Greek National School of Public Administration;
- Chairman of the Steering Committee of the Joint Activity on Urban Management, Organization of Economic Cooperation and Development (OECD);
- Chairman of the Greek National Council of Spatial Planning and Sustainable Development;

L. C. Wassenhoven, *The History of Town Planning in a Dependent State*,
https://doi.org/10.1007/978-3-031-96958-4

- Vice-Chairman of the Greek State Scholarships Foundation;
- Fellow scholar, University of Melbourne;
- Member of the Council of the Crete Technical University;
- Member of the Board of the Public Company for the Redevelopment of the former "Hellinikon" International Airport of Athens;
- Member of the Greek Committee of Experts for the Codification of Urban and Regional Planning Legislation.

He has coordinated a large number of research projects of the National Technical University of Athens, and has written many articles and the books listed below.

Books

In English:

1. The Ancestry of Regional Spatial Planning: A planner's look at history (Springer Publishers 2019).
2. Compromise planning: A theoretical approach from a distant corner of Europe (Springer Publishers 2022).
3. Spatial planning as institutional design: European countries amid economic crisis and austerity (Edward Elgar Publishers 2024).
4. The EU Compendium of Spatial Planning Systems and Policies: Greece (European Commission Regional Development Studies 2000).

In Greek:

1. Territorial Covernance: Theory, European experience and the case of Greece (Kritiki Publishers 2010).
2. Maritime Spatial Planning: Europe and Greece (Crete University Press 2017).
3. Hellenic… Passion: The development of the former Hellinikon airport (Sakkoulas Publishers 2018).
4. Putting our Country in order: History of regional spatial planning in Greece after World War II (Kritiki Publishers 2022).
5. Looking for the Future of European Space: Territorial cohesion, spatial planning, cities and climate change (Propobos Publishers 2023).
6. Theories and Praxis of Spatial Planning: From the invisible to the visible (Propobos Publishers 2024).
7. History of Town Planning in Greece, 1821-1940: Institutions, plans, historical framework, international milieu, foreign influence (Sakkoulas Publishers 2025) (in co-operation with V. Angelidou).
8. Town Planning in a Maelstrom: Fifty years from the 1975 Constitution to the Local Urban Plans (Sakkoulas Publishers 2025).[1]

[1] The books in Greek numbered 7 and 8 were published in a series of publications coordinated jointly by the Sakkoulas Publishers and the Town Planning Research Laboratory of the National Technical University of Athens.

Index

L. C. Wassenhoven, *The History of Town Planning in a Dependent State*, https://doi.org/10.1007/978-3-031-96958-4

The manufacturer's authorised representative in the EU is Springer Nature Customer Service Centre GmbH, Europaplatz 3, 69115 Heidelberg, Germany. If you have any concerns regarding our products, please contact ProductSafety@springernature.com

Printed and bound by CPI Group (UK) Ltd, Croydon, CR0 4YY
01/10/2026
02225971-0001